Jan Henrik Schmidt

# An efficient solution procedure for elastohydrodynamic contact problems considering structural dynamics

**Karlsruher Institut für Technologie**
**Schriftenreihe des Instituts für Technische Mechanik**

Band 33

Eine Übersicht aller bisher in dieser Schriftenreihe erschienenen Bände finden Sie am Ende des Buchs.

# An efficient solution procedure for elastohydrodynamic contact problems considering structural dynamics

by
Jan Henrik Schmidt

KIT Scientific Publishing

Dissertation, Karlsruher Institut für Technologie
KIT-Fakultät für Maschinenbau

Tag der mündlichen Prüfung: 21. November 2018
Referenten: Prof. Dr.-Ing. Wolfgang Seemann, Prof. Dr. Nicolas Fillot

**Impressum**

KIT Scientific Publishing

Karlsruher Institut für Technologie (KIT)
KIT Scientific Publishing
Straße am Forum 2
D-76131 Karlsruhe

KIT Scientific Publishing is a registered trademark
of Karlsruhe Institute of Technology.
Reprint using the book cover is not allowed.

www.ksp.kit.edu



Print on Demand 2018 – Gedruckt auf FSC-zertifiziertem Papier

ISSN 1614-3914
ISBN 978-3-7315-0872-4
DOI 10.5445/KSP/1000088235

# An efficient solution procedure for elastohydrodynamic contact problems considering structural dynamics

Zur Erlangung des akademischen Grades

**Doktor der Ingenieurwissenschaften**

der Fakultät für Maschinenbau
Karlsruher Institut für Technologie (KIT)

genehmigte
**Dissertation**
von

Dipl.-Ing. Jan Henrik Schmidt
aus Hannover

| | |
|---|---|
| Tag der mündlichen Prüfung: | 21. November 2018 |
| Hauptreferent: | Prof. Dr.-Ing. Wolfgang Seemann |
| Korreferent: | Prof. Dr. Nicolas Fillot |

# Acknowledgements

During my studies, I experienced countless acts of kindness from many people. These acts, academic and non-academic in nature, profoundly influenced this work and my life. For this, I am truly grateful.

I would like to acknowledge some individuals. As a comprehensive list would fill multiple pages, I cannot thank each and every one of them. Still, I am grateful for the kindness I was shown by every single one, whether listed here or not.

First of all, I would like to express my sincere gratitude to my advisor, Prof. Dr.-Ing. Wolfgang Seemann, who guided me throughout my Ph.D. studies and significantly improved my work and analytical skills through many discussions, good advice, and his attention to detail.

I am thankful to Prof. Dr. Nicolas Fillot for his interest in this work and the discussions and comments that helped improve this dissertation.

I am indebted to Dr.-Ing. Sergey Solovyev for his trust, guidance, and support. He influenced this work significantly through countless discussions, great dedication, and, last but not least, many shared laughs.

I would like to thank the Robert Bosch GmbH for funding this research and providing a great working environment. Both at Bosch and at the Institute of Engineering Mechanics in Karlsruhe, I had the pleasure of working with and learning from many wonderful colleagues. I am particularly thankful to Dr.-Ing. Daniel Maier, M.Sc. Johannes Bette, and Dr.-Ing. Oliver Stump, all of whom were integral to this journey—as both colleagues and friends.

I also would like to thank Prof. Dr.-Ing. Hartmut Hetzler for encouraging and productive conversations and Prof. Dr.-Ing. habil. Alexander Fidlin for motivating me with challenging questions and comments. I am indebted to Prof. Dr.-Ing. habil. Stefan Hartmann and Dr.-Ing. Steffen Rothe for sparking my interest in the study and research of computational mechanics.

I would like to acknowledge the students who trusted me to guide them on their master's theses. I am grateful for their time, dedication, and for challenging me with many difficult questions.

My deepest gratitude goes to my friends and family. Their continued support, their encouragement during difficult times, and their welcome distractions made this work possible. Above all, I would like to thank my family for their unconditional love.

# Abstract

Conventional elastohydrodynamic (EHD) contact models usually approximate deformation using half-space theory while neglecting structural inertia effects. The conventional modeling approach yields good results for compact bodies. However, it is not suitable for more complex bodies where structural inertia effects influence the contact behavior. This work proposes an efficient solution procedure which considers structural inertia effects and is applicable to arbitrary geometries, thus, providing a superior solution to the EHD contact problem.

The contact bodies are modeled by specially adapted reduced finite element models. Singly diagonal implicit Runge-Kutta (SDIRK) methods are used for adaptive time integration. A fluid-structure coupling is presented to combine the structural model and the nonlinear Reynolds Equation using a monolithic coupling approach. Finally, a reduced order model of the complete nonlinear coupled problem is constructed.

The solution procedure is studied on the basis of familiar example problems from EHD literature, namely the shock loading of an EHD contact and the lubricated impact problem. To investigate the extended structural model, the impact of a valve-like geometry is examined for both a dry contact and a lubricated contact. The proposed solution procedure is able to capture effects in the contact behavior not captured by the conventional modeling approach needing little additional computational cost.

# Kurzfassung

Bei der Modellierung von elastohydrodynamischen (EHD) Kontakten wird die Verformung häufig mittels der Halbraumtheorie approximiert und das dynamische Verhalten der Kontaktkörper oftmals vernachlässigt. Dieser konventionelle Modellierungsansatz liefert gute Ergebnisse für kompakte Körper, ist jedoch nicht für komplexere Geometrien geeignet, bei welchen das Kontaktverhalten durch Trägheitseffekte beeinflusst wird. In dieser Arbeit wird ein effizientes Lösungsverfahren vorgestellt, welches das dynamische Verhalten der Kontaktkörper berücksichtigt und auf beliebige Geometrien anwendbar ist. Die verbesserte Modellierung ermöglicht die Verwendbarkeit bei komplexeren Problemstellungen.

Die Kontaktkörper werden durch speziell angepasste, reduzierte Finite-Elemente-Modelle dargestellt. Zur adaptiven Zeitintegration werden einfach diagonal implizite Runge-Kutta Verfahren verwendet. Mittels einer Fluid-Struktur-Kopplung wird das erweiterte Strukturmodell mit der nichtlinearen Reynolds-Gleichung gekoppelt. Das monolithisch gekoppelte Problem wird mit dem Newton-Verfahren gelöst. Darüber hinaus wird ein reduziertes Modell des gesamten nichtlinearen Problems erstellt.

Das entwickelte Lösungsverfahren wird anhand bekannter Beispiele aus der EHD Literatur untersucht. Dabei wird die schlagartige Belastung eines EHD-Kontakts sowie das Stoßproblem betrachtet. Weiterhin wird das erweiterte Strukturmodell anhand einer ventilähnlichen Geometrie sowohl für einen trockenen als auch für einen geschmierten Kontakt untersucht. Das erweiterte EHD-Kontaktmodell kann Effekte im Kontaktverhalten mit geringem zusätzlichem Rechenaufwand abbilden, die mittels des konventionellen Modellierungsansatzes nicht abbildbar sind.

# Contents

# Nomenclature

## Remarks on the notation

To improve the understanding of relations between different quantities, scalars, geometrical vectors, column matrices and matrices are visualized as follows:

- Scalars $a$,
- Geometrical vectors $\vec{a}$,
- Column matrices $\boldsymbol{a}$, $\boldsymbol{A}$, and
- Matrices $\mathbf{A}$.

To identify selective rows and columns of a matrix, a programming-like notation is used. For example, a submatrix, consisting of the first $n_\mathrm{B}$ rows and the first $m_\mathrm{B}$ columns of the matrix $\mathbf{A} \in \mathbb{R}^{n_\mathrm{A} \times m_A}$, will be denoted as

$$\mathbf{B} = \mathbf{A}(1 : n_\mathrm{B},\ 1 : m_\mathrm{B}) = \begin{bmatrix} a_{11} & \cdots & a_{1m_\mathrm{B}} \\ \vdots & \ddots & \vdots \\ a_{n_\mathrm{B}1} & \cdots & a_{n_\mathrm{B}m_\mathrm{B}} \end{bmatrix} \in \mathbb{R}^{n_\mathrm{B} \times m_\mathrm{B}},$$

where $a_{ij}$ is the element in the $i$-th row and $j$-th column of the matrix $\mathbf{A}$.

## Scalars

| | |
|---|---|
| $\alpha_\eta$ | Pressure-viscosity-coefficient |
| $\beta_\eta$ | Pressure-viscosity-index |
| $\gamma_\mathrm{pen}$ | Penalty factor |
| $\delta$ | Gap displacement |
| $\bar{\delta}$ | Dimensionless gap displacement |
| $\epsilon_\mathrm{a}$, $\epsilon_\mathrm{r}$ | Absolute and relative tolerance of the step size controller |

| | |
|---|---|
| $\zeta$ | Damping ratio |
| $\eta$ | Viscosity |
| $\bar{\eta}$ | Dimensionless viscosity |
| $\eta_0$ | Viscosity at ambient pressure |
| $\nu$ | Poisson's ratio |
| $\xi$ | Local coordinate |
| $\xi_{i_Q}$ | Local coordinate of a quadrature point |
| $\rho$ | Density |
| $\bar{\rho}$ | Dimensionless density |
| $\rho_0$ | Density at ambient pressure |
| $\chi_e$ | Transformation from local to global coordinates |
| $\omega^{(m)}$ | Under-relaxation factor for the $m$-th Newton-Raphson iteration |
| $\hat{b}_k$ | Weight coefficient of an embedded Runge-Kutta method |
| $a_{kl}, b_k, c_k$ | Weight coefficient of a Runge-Kutta method |
| $b$ | Hertzian half-width |
| $E$ | Young's modulus |
| $E'$ | Effective Young's modulus |
| $err_{\text{max}}$ | Maximum absolute deviation, see Eq. (A.3) |
| $f_{\text{c}}$ | Cut-off frequency |
| $F_c$ | Central force, $F_c = F(t, X \approx 0)$ |
| $f_{\text{min}}, f_{\text{max}}$ | Parameters of the time step size controller |
| $f_{\text{safety}}$ | Stabilization parameter of the time step size controller |
| $g$ | Acceleration of gravity |

| | |
|---|---|
| $h$ | Gap height |
| $H$ | Dimensionless gap height |
| $h_0$ | Initial height |
| $h_c$ | Central gap height, $h_\mathrm{c} = h(t, X \approx 0)$ |
| $h_\mathrm{min}$ | Minimum gap height, $h_\mathrm{min}(t) = \min_{X \in \Omega_\mathrm{F}} (h(t, X))$ |
| $i_\mathrm{GR}$ | Grid ratio |
| $l$ | Loss of information |
| $l_\mathrm{max}$ | Maximum tolerable loss of information |
| $m_0$ | Mass |
| $\widetilde{N}_p$ | Number of points for the nonlinear function evaluation |
| $\hat{N}_p$ | Number of points for the state variable evaluation |
| $n_\mathrm{cm}$ | Number of constraint modes |
| $\widetilde{n}_\delta$ | Number of basis functions for the gap displacement |
| $n_\mathrm{en}$ | Number of nodes per element |
| $n_\mathrm{F}$ | Number of fluid nodes |
| $n_\mathrm{nm}$ | Number of normal modes |
| $n_p$ | Number of unknown pressures |
| $\widetilde{n}_p$ | Number of basis functions for the pressure |
| $n_\mathrm{Q}$ | Number of quadrature points |
| $n_q$ | Number of degrees of freedom in modal space |
| $n_\mathrm{s}$ | Number of snapshots |
| $n_\mathrm{S}$ | Number of structure nodes |
| $n_t$ | Number of time steps |

| | |
|---|---|
| $n_u$ | Number of displacement degrees of freedom |
| $\widetilde{n}_u$ | Number of basis functions for the displacement |
| $n_y$ | Number of state variables in the state-space representation |
| $n_z$ | Number of unknowns in the full system |
| $\widetilde{n}_z$ | Number of basis functions of the reduced system |
| $p$ | Pressure |
| $P$ | Dimensionless pressure |
| $p, \hat{p}$ | Order of a Runge-Kutta method |
| $p_c$ | Central pressure, $p_c = p(t, X \approx 0)$ |
| $p_H$ | Hertzian pressure |
| $R$ | Radius of curvature |
| $R_{\text{red}}$ | Reduced radius of curvature |
| $relerr$ | Relative deviation over time, see Eq. (A.4) |
| $relerr_{\text{max}}$ | Maximum relative deviation, see Eq. (A.5) |
| $relerr_{\text{mean}}$ | Mean relative deviation, see Eq. (A.2) |
| $t$ | Time |
| $T$ | Dimensionless time |
| $\Delta t_j$ | Time step size of the $j$-th time step |
| $\Delta T_j$ | Dimensionless time step size of the $j$-th time step |
| $t_0$ | Initial time |
| $v_0$ | Initial impact velocity |
| $w$ | Load |
| $\bar{w}$ | Dimensionless load |

| | |
|---|---|
| $w_{i_Q}$ | Weight coefficient of Gauss quadrature |
| $x, y, z$ | Spatial coordinates |
| $\Delta x^F$ | Size of a fluid element |
| $\Delta x^S$ | Size of a structure element |
| $X$ | Dimensionless spatial coordinate |

## Vectors and matrices

| | |
|---|---|
| $\boldsymbol{\varphi}$ | Eigenvector |
| $\bar{\boldsymbol{\Phi}}$ | Modal matrix |
| $\hat{\boldsymbol{\varphi}}$ | Mass normalized eigenvectors |
| $\boldsymbol{\Phi}_{\mathrm{cm}}$ | Constraint mode matrix |
| $\boldsymbol{\Phi}_{\mathrm{nm}}$ | Normal mode matrix |
| $\boldsymbol{\delta}$ | Gap displacement vector |
| $\bar{\boldsymbol{\delta}}$ | Dimensionless gap displacement vector |
| $\mathbf{A}, \mathbf{B}, \mathbf{C}, \mathbf{D}$ | Matrices of the state-space representation |
| $\overline{\mathbf{D}}$ | Modal damping matrix |
| $\boldsymbol{F}$ | Force vector |
| $\boldsymbol{f}$ | Discrete system function |
| $\tilde{\boldsymbol{F}}$ | Reduced force vector |
| $\mathbf{G}_{\mathrm{FQ}}$ | Interpolation matrix—from fluid grid to quadrature point |
| $\mathbf{G}_{\mathrm{int}}$ | Pressure integration matrix |
| $\mathbf{G}_{\mathrm{QS}}$ | Integration matrix—from quadrature point to structure grid |
| $\mathbf{G}_{\mathrm{SF}}$ | Interpolation matrix—from structure to fluid grid |
| $\boldsymbol{h}$ | Gap height vector |

| | |
|---|---|
| $\boldsymbol{H}$ | Dimensionless gap height vector |
| $\mathbf{J}_f$ | Jacobian matrix of the system function |
| $\mathbf{K}$ | Stiffness matrix |
| $\tilde{\mathbf{K}}$ | Reduced stiffness matrix |
| $\overline{\mathbf{K}}$ | Reduced stiffness matrix in modal space |
| $\mathbf{M}$ | Mass matrix |
| $\tilde{\mathbf{M}}$ | Reduced mass matrix |
| $\overline{\mathbf{M}}$ | Reduced mass matrix in modal space |
| $\mathbf{N}$ | Matrix of shape functions |
| $\boldsymbol{p}$ | Pressure vector |
| $\boldsymbol{P}$ | Dimensionless pressure vector |
| $\boldsymbol{q}$ | Generalized coordinates |
| $\boldsymbol{S}^{jk}$ | Starting vector of a Runge-Kutta method |
| $\boldsymbol{u}$ | Displacement vector |
| $\boldsymbol{U}$ | Dimensionless displacement vector |
| $\tilde{\boldsymbol{u}}$ | Reduced displacements |
| $\boldsymbol{u}^{\mathrm{F}}$ | Displacements on the fluid grid |
| $\boldsymbol{u}^{\mathrm{S}}$ | Displacements on the structure grid |
| $\mathbf{V}_\delta$ | Projection matrix for the gap displacement |
| $\mathbf{V}_p$ | Projection matrix for the pressure |
| $\mathbf{V}_u$ | Projection matrix for the displacement |
| $\mathbf{V}_z$ | Projection matrix for the full system |
| $\boldsymbol{x}^{\mathrm{F}}$ | Fluid grid point coordinates |

| | |
|---|---|
| $\boldsymbol{x}^{\mathrm{S}}$ | Structure grid point coordinates |
| $\boldsymbol{y}$ | State variable |
| $\boldsymbol{Y}^{jk}$ | Stage value of a Runge-Kutta method |
| $z$ | State vector of the full system |
| $\tilde{z}$ | State vector of the reduced system |

# 1 Introduction

The demands on modern machine elements are continuously increasing. While machine elements are to become cheaper, smaller and lighter, the expectations towards functionality and durability are not to be compromised. In addition to satisfying basic requirements, other desired aspects, e.g., good acoustic performance, play an essential role. In order to meet all these requirements, a precise understanding of the machine elements and the interaction of the components is necessary. Numerical studies are increasingly used in the product development process in order to gain a broader and deeper understanding of interrelationships. The focus of recent studies is to take into account more and more physical effects in the simulation models, often resulting in coupled problems. The numerical solution of the coupled problems, however, can quickly become time-consuming. For this reason, simulation techniques are continuously developed and improved to enable a fast and stable numerical solution of the models.

The elastohydrodynamic (EHD) contact problem is a coupled problem that typically occurs in non-conformal lubricated contacts. Such contacts exist, for example, in ball bearings or gears. The EHD contact problem is characterized by very high viscosities and elastic deformations that are greater than the gap height. To predict the behavior of an EHD contact, the solid's behavior and the fluid's behavior must be considered at the same time. In EHD contacts the fluid flow is typically modeled using the nonlinear Reynolds Equation. The fluid's behavior is modeled by viscosity and compressibility relations, while the solid's deformation is commonly modeled using the half-space approach. The structural dynamic behavior of the bodies is generally not considered in the conventional EHD contact model.

In some applications, e.g., valve-seat/valve-needle contacts, the conventional EHD model does not suffice to meet the above requirements. Such applications have more complex geometries and boundary conditions that influence the contact behavior and must be considered in the investigations. To meet the demands of these applications, model extensions are necessary. However, model extensions are often accompanied by an extension of the computational time. Hence, efficient solution procedures are needed to keep the additional computational cost low.

This work aims to extend the structural modeling of the transient EHD contact model using modern numerical methods. The resulting model shall be applicable to a wide range of machine elements while keeping the computational time low. The solution of the lubricated impact problem is an interdisciplinary problem. To this end, a brief survey of relevant literature is presented in the following.

## 1.1 Literature review

The study of lubricated contact problems is a multi-disciplinary field with application in various domains. In this work, the focus lies on efficient solution procedures for non-steady-state EHD contact problems. To this end, first, a short historical review of the EHD contact problem is given. Next, developments of numerical solution procedures in this field are summarized. This work focuses on the lubricated impact problem, a review of this problem concludes the literature review.

#### Historical developments of the EHD contact problem

Following the experimental studies of Petrov and Tower on railway axle bearings, Osborne Reynolds [110] laid the foundation for the calculation of lubricated contacts in 1886 by deriving a partial differential equation to describe the flow of Newtonian fluids in narrow gaps. This equation is referred to as Reynolds Equation. The improved theoretical understanding resulting from Reynolds' studies helped in the development of plain bearing technology. However, in the study of gears, results obtained solely with Reynolds Equation yield very small gap heights, which were not in agreement with the experimentally observed longevity of gears. It became clear that the observed behavior was related to an increase in viscosity and local elastic deformations, which resulted from high local pressures in the contact of gear teeth. However, theoretical investigations remained inconclusive. A breakthrough was finally achieved by Ertel [41], whose work was published by Grubin [51] in 1949. Ertel and Grubin managed to incorporate both the increase in viscosity and the local elastic deformations in one theory. The results of these investigations provided significantly larger gap heights than previous studies. Consequently, the longevity of gears could finally be explained theoretically. The findings by Ertel and Grubin mark the cornerstone of the elastohydrodynamic (EHD) contact theory.

A detailed overview of the historical developments of the EHD contact problem can be found in [33, 36, 90, 117].

#### Developments of the numerical solution procedure of EHD contact problems

First numerical solutions of an EHD contact problem were presented by Petrusevich [104] in 1951. Petrusevich's solutions simultaneously satisfied Reynolds Equation and the elasticity relations. His solutions were the first to show characteristic EHD contact features, such as the pressure peak, which forms at the outlet of the contact. This narrow peak is also referred to as the Petrusevich spike. Further remarkable results were published by Dowson and Higginson [34] in 1959. They presented numerical solutions for a broader range of parameters and derived empirical regression equations from the solutions. The resulting equations could express the line contact problem in terms of three dimensionless groups. These groups were often used in the following years to gain an understanding of EHD contact behavior.

Dowson and Higginson [34] used the inverse method to compute their numerical solutions. The inverse method uses an inverted and integrated Reynolds Equation to compute a gap height profile from a given pressure distribution. The same pressure distribution is then used to compute a second gap height profile from the elasticity relations. In multiple iterations, the pressure distribution is manually modified until the two gap height solutions coincide. Two disadvantages of this approach are that an accurate initial estimation of the pressure distribution is needed, which is often available only for highly loaded contacts, and that the modifications of the pressure distribution are carried out by hand, which requires skill and insight into the contact behavior. Nonetheless, the inverse method was successfully adapted to compute solutions for the point contact problem by Evans and Snidle [43] in 1982.

Another solution approach is the direct method. The direct method uses Reynolds Equation to compute a pressure distribution from a given gap height profile. The resulting pressure distribution is then used to compute a new gap height profile from the elasticity relations. This procedure is repeated until convergence is reached. It was first applied to the line contact problem by Stephenson and Osterle [119] in 1962 and to the point contact problem by Ranger et al. [107] in 1975. Using this approach, the iterations were usually performed using Gauss-Seidel method [20, 64, 65]. Disadvantages are the bad convergence for highly loaded cases, as well as a large number of required iterations. Hence, long computational times were needed and convergence was rarely archived for highly loaded cases.

There have been two main approaches to overcome these problems. One approach is to strengthen the coupling between Reynolds Equation and the elasticity relations, i.e., a monolithic coupling approach. Another approach is to keep the partitioned coupling and solve the problem using multigrid methods. The multigrid method, which is based on the Gauss-Seidel method, was first introduced to the EHD contact problem by Lubrecht [87–89] in 1986. The multigrid method makes use of the difference in convergence velocity of errors on different grid refinement levels. The computation of the elastic deformations dominated the computational time of the original method. The multilevel multi-integration approach [11] improved the efficiency of the computation of the elastic deformations. Venner [126] introduced further improvements in 1991. His distributive relaxation schemes improved the stability of the procedure. Venner and Lubrecht outlined the implementation of multigrid methods in lubricated contact problems in their book [128] in 2000. The ready availability of the method, paired with its computational efficiency and stability for highly loaded cases, make this method a popular choice for the solution of EHD contact problems until today.

A stronger coupling results from solving the problem as one system of equations, resulting in the simultaneous computation of one set of solutions for the pressure and the gap height. This approach was applied to EHD contact problems by Okamura [101] in 1982 and later by Oh

[100] and Houpert and Hamrock [74]. The coupling improves the stability of the numerical procedures, which allows the computation of highly loaded cases. The resulting system of equations is commonly solved using the Newton-Raphson method, which requires an inversion of the Jacobian matrix in every iteration. Unfortunately, the Jacobian matrix relating to the elastic terms is fully populated. The inversion of a fully populated matrix of size $n \times n$ has complexity $O(n^3)$. Therefore, the computational effort of this approach becomes high for a large number of grid points. To overcome this problem, Evans and Hughes [42] and Hughes et al. [75] proposed the differential deflection method. This method enables the construction of a banded matrix relating the pressure to the deformation by making use of the fast decay of the second differential of the deformation with pressure. The method was further extended by Holmes et al. [72, 73] and applied to transient problems. However, the combination of the complex implementation with remaining computational drawbacks for point contact problems has likely hindered the spreading of the monolithic coupling approach in the context of EHD contact problems. More recently, Habchi et al. [55, 60, 61] proposed the usage of the finite element method to solve Reynolds Equation as well as the elasticity relations. Discretization using finite elements yields a sparse matrix, thus, overcoming the issue of the densely populated matrix, which was an obstacle in half-space approach based methods. Additionally, they proposed using the penalty method for the treatment of the free boundary cavitation problem [137], which eases the implementation of a monolithic coupling approach. Using finite elements for the discretization of the solids creates a new challenge. Namely, that many degrees of freedom, which have no direct influence on the behavior in the contact region, are introduced into the system of equations. Nevertheless, Habchi et al. showed that the computational complexity of their proposed procedure is comparable to existing solvers which use the multigrid method. The improved computational efficiency results from the faster convergence of the monolithic coupling approach.

In recent years efforts have been made to decrease further the computational time of EHD contact problems using model order reduction techniques. First successful attempts were presented by Habchi and Issa [57, 58] and by Habchi et al. [56]. They applied their so-called EHD-basis technique to reduce the linear part of the EHD contact model, i.e., the elasticity relations. As reduction basis, they used elastic deflections, which were computed in advance using the full EHD contact model. Therefore, this reduction approach requires an offline phase, i.e., a phase were training solutions are computed in advance using the full model. Maier [93] reduced the EHD contact problem using two different model order reduction approaches. One approach is based on the proper orthogonal decomposition method and a subsequent system approximation [95]. A second approach is based on the trajectory piece-wise linear method [94, 108, 109]. The second approach, however, can only be applied to transient problems. Maier et al. showed tremendous decreases of the online computational time resulting from these approaches. On the other hand, these approaches also require an upfront offline phase in which the reduced order models are constructed based on solutions of the full model. Furthermore, the functionality of the reduced order models is then only given in the parameter space for which the reduced order model has been constructed, thus, limiting the area of application of the reduced order model.

A very recent study in the context of model order reduction techniques applied to EHD contact problems was presented by Habchi and Issa [59] in 2017. As in their previous approaches, Habchi and Issa only reduced the linear part of the model. However, different from previous approaches, they reduced the elasticity relations using static condensation. Static condensation is often referred to as the Guyan reduction [54]. This method consists only of matrix operations and is therefore exact for static problems. However, this approach is not suitable for structural dynamics problems [106]. The method has the advantage that it depends solely on the stiffness matrix. Hence, no computations using the full model are required for the construction of the reduced order model.

Out of the many works focusing on the development of efficient numerical techniques to solve the EHD contact problem few have focused on the time integration procedure. The temporal term of Reynolds Equation is commonly discretized in time by the Backward Euler method [38, 48, 85, 100, 126, 127] or the second-order finite difference method [94, 124, 134]. Venner et al. [124] and Wijnant [134] pointed out that the damping introduced by the Backward Euler method noticeably changes the contact behavior and proposed the usage of the second-order finite difference method. The second-order finite difference method is widely used today. Regarding the time step size the first solutions of the EHD contact problem using adaptive time stepping were presented by Goodyer [47] in 2001. Another approach was presented by Watremetz et al. [132]. Nonetheless, the use of a fixed time step size remains a typical choice until today [129].

### Developments of the lubricated impact problem

The first studies of the EHD contact problem mostly neglected the pressure resulting from the squeeze term, i.e., the temporal term of Reynolds Equation was neglected. One of the first theoretical researches of EHD contacts under impact motion was carried out by Christensen [21] in 1962. He presented theoretical and experimental results for two normally approaching cylinders separated by a lubricant. His numerical analysis considered an incompressible lubricant with exponentially varying viscosity and an elastically deforming solid. However, he neglected the contribution of the elastic deformations to the temporal term of Reynolds Equation. He presented two important observations. Namely, that the maximum pressure in the lubricated case is larger than in the dry contact case. Moreover, that in the lubricated case the minimum film thickness does not occur at the center of the contact. His experimental results showed larger plastic deformations in the contact region for the lubricated contact case than for the dry contact case. These observations validate his theoretical results and are in agreement with the observations of Bowden and Tabor [10]. Direct evidence of the formation of a central dimple was presented in 1967 by Dowson and Jones [37]. They studied the deformation of the solids around the highly viscous lubricant using optical interferometry.

In 1970, Christensen [22] published another paper on the normal approach. In this paper, he studied an axisymmetrically modeled spherical body. The results confirmed his observations for the line contact problem from his first paper [21]. However, he did not adequately consider the elastic deformations in the temporal term of Reynolds Equation in either approach. In the same year, further theoretical studies of the normal approach of two cylinders separated by a lubricant were presented by Herrebrugh [69]. Following Christensen, he assumed a constant approach velocity of the surfaces and did not consider elastic deformations in the temporal term of Reynolds Equation. He notes however that this effect should be considered for small gap heights.

In 1973, Lee and Cheng [86] extended the previous modeling approaches by considering the elastic deformations in the temporal term of Reynolds Equation. They were able to reach convergence for smaller gap heights and noted the relevance of the local elastic deformations. Similar to previous studies, they assumed the cylinders to be weightless, thus, not solving the equation of motion. Conway and Lee [24], on the other hand, considered the equation of motion but neglected the elastic deformations. They highlighted the role of the piezoviscosity in the formation of the central dimple.

Interesting results of the lubricated impact problem were presented by Safa and Gohar [114] in 1986. Safa and Gohar dropped a steel sphere on a lubricated glass plate. They measured the pressure-time trace in the contact region using manganin pressure transducers. Their experimental results showed a secondary pressure peak in the contact center at the end of the impact-rebound process. Besides the observation of the secondary pressure peak, they noted that even without replenishment of the oil droplet no noticeable changes in the pressure profile occur for the case of multiple impacts. This observation indicates that the pressure buildup begins at very small gap heights.

The first transient solutions of a sphere dropping on a lubricated plate were presented by Yang and Wen [138] in 1991. They considered the mass of the sphere as well as the elastic deformations in the temporal term of Reynolds Equation. The equation of motion was solved using the explicit Euler method and Reynolds Equation was solved using the Backward Euler method. Results for the first part of the impact process were presented, leaving the question about the secondary pressure peak unanswered. They highlighted the significance of the squeeze effect only for small gap heights, thus, confirming the observations of Safa and Gohar in this regard.

The first numerical results of the complete impact and rebound of an elastic sphere on a lubricated plate were presented at the 6th Nordic Symposium by Dowson and Wang [38] and by Larsson and Höglund [85]. Both studies showed the secondary pressure peak, which was experimentally observed by Safa and Gohar [114].

Further studies investigating EHD contacts with rapid normal motion followed. Particular interest was given to the effects of loading speed and/or initial gap height. In this context, the occurrence of noncentral dimples was observed experimentally by Sakamoto et al. [115]. Numerical results were presented by Guo et al. [52] and by Kaneta et al. [80, 82] and Nishikawa et al. [98].

In further studies, Kaneta et al. [81] investigated the influence of surface features while Wang et al. [130, 131] studied the influence of thermal effects. Efforts were made by Wang, Venner and Lubrecht to obtain simple equations for the prediction of transient EHD contact behavior for line contacts [129] and for point contacts [125]. While these relations are helpful to gain an understanding of the problem, they do not give information about the spatial resolution and transient behavior of the pressure and film thickness.

In the studies listed above, the numerical solution procedure and the mathematical model of the EHD contact problem have remained mostly unchanged. For the solution of the coupled problem, the multigrid procedure as given by [128] was used. Hence, equations were coupled using a partitioned coupling approach and Reynolds Equation was discretized using finite differences on an equidistant grid in space and time. The mathematical model describing the behavior of the solids have remained unchanged to that used in the early 1990s, i.e., the elastic deformations are approximated using the half-space approach and the inertia of the solids is considered by the equation of motion of a point mass.

## 1.2 Objective and outline

From the review of previous research, it becomes clear that the multigrid procedure remains the most popular solution procedure until today. Similarly, the elastic half-space approach has almost without exception been used for the approximation of the elastic deformations in EHD contact problems. Structural inertia is, if at all, considered by the equation of motion of a point mass. This modeling approach does not account for the dynamic behavior of elastic bodies with complex geometries. In certain machine elements, the dynamic behavior of the contact bodies influences the contact behavior. Hence, the application of conventional EHD contact models for such applications can lead to unsatisfactory results. To overcome this drawback, efforts have been made to construct viscous force-damper elements from EHD contact simulations and to consider these in the context of a structural dynamic simulation. However, to the author's knowledge, no modeling approach exists which computes the structural behavior and the fluid contact behavior in one fully coupled system of equations.

The objective of this work is to develop an efficient numerical solution procedure for the lubricated impact problem, incorporating the dynamic behavior of the contact bodies. To this end, the conventional modeling of the EHD contact problem is extended with respect to the structural model. The additional computational effort to solve the extended model shall be low. To this end, modern numerical techniques are applied, such that the computational time of the EHD contact problem using the extended modeling approach is comparable to the computational time of the less complex conventional EHD contact model.

## Outline

The second chapter illustrates the modeling of the lubricated impact problem and outlines numerical methods used to solve the resulting model. First, the fluid model, as it can be found in the EHD literature, is presented. The case of the line contact and the axisymmetric point contact is shown. Then, the nondimensionalization and discretization of Reynolds Equation using finite differences is presented. Next, the structural model is presented. It is based on the finite element formulation and coupled with the fluid model via the gap height. Section 2.3 summarizes the model order reduction procedures used to reduce the dimensions of the finite element model. Starting from a projection onto a lower dimensional subspace via modal truncation and static condensation, the method of component mode synthesis is motivated. Finally, Section 2.4 presents the adaptive time integration using singly diagonal implicit Runge-Kutta method. Here, the used time integration scheme, the basic algorithm of the time integration method, and the time step size control is shown.

Based on the equations and methods of Chapter 2, Chapter 3 describes the developed solution procedure of the coupled problem. The first part of the chapter focuses on the construction of the structural model and its coupling into the lubricated impact problem. In this context, the reduced structural model is decoupled through a modal transformation. Next, methods to minimize spurious oscillations are presented, which can be observed in the context of reduced order models. An efficient procedure for the coupling of the two bodies with the fluid is presented subsequently. Due to the used procedures, the resulting equation system has the same size as the equation system of conventional EHD models. The fully coupled equation system is solved in a monolithic approach using the Newton-Raphson method. Finally, Section 3.7 presents the model order reduction of the entire nonlinear system. To this end, the coupled problem is first projected onto a subspace, before, in a second step, the size of the reduced system functions is further reduced by a system approximation.

In Chapter 4 the presented solution procedure is applied to example problems. Three different problems are considered, which allows for the separate investigation of different parts of the solution procedure. The first example problem is the shock loading of a lubricated contact. For this problem, a conventional EHD model is used. Based on this problem, the proposed time integration procedure is studied. Additionally, an extension of the time step size control based on the monitoring of parameter changes is presented.

The second example problem is the dry contact of a valve-like geometry. Using this example problem, the extended structural model and the methods for dealing with spurious oscillations in reduced structural models are investigated.

The third example problem is the lubricated impact problem. For this example problem, the entire extended EHD model is used. To validate the model, a comparison with results from literature is made. Moreover, the influence of the extended EHD model is illustrated by varying the material parameters of a valve-like geometry. The numerical aspects of the model are briefly examined. Finally, the application of multiple impacts is examined. Such problems can occur, for example, in the calculation of wear processes. This study allows the investigation of the model order reduction of the entire system. The proposed procedure automatically creates local reduced order models to decrease the computational effort of problems with recurrent processes.

Chapter 5 concludes this thesis and motivates further work.

# 2 Theory

This chapter summarizes the equations used to model the lubricated contact problem and describes numerical methods to solve this problem. First, the fluid model is presented. The line contact and the axisymmetric point contact model are shown. Next, Section 2.2 outlines the structural model. In contrast to conventional approaches, the structural bodies are discretized using finite elements. The resulting finite element model is reduced using model order reduction procedures. Section 2.3 presents model order reduction techniques with an emphasis on structural dynamics. Finally, Section 2.4 outlines the adaptive time integration procedure using singly diagonal implicit Runge-Kutta (SDIRK) methods.

## 2.1 Fluid model

Based on simplified Navier-Stokes equations, Osborne Reynolds derived in 1886 an equation to describe slow viscous flow in narrow gaps [110]. In this context, he assumed—among other things—that viscous terms are predominant and inertial and volume forces are negligible in comparison. Furthermore, the pressure is assumed to be constant in the film thickness direction and the fluid to behave Newtonian. These assumptions significantly reduce the complexity of the full Navier-Stokes equations, while yielding good results for lubricated contact applications, which makes the Reynolds Equation a popular choice to model the fluid flow in lubricated contact problems.

The following subsections summarize the fluid model of the pure impact EHD contact problem. The mathematical model for the rolling/sliding EHD contact problem is summarized in Appendix B.

The theory presented in the following is based on the works [5, 66, 123, 125, 128, 129], which provide a more detailed overview.

### 2.1.1 Reynolds Equation

The isothermal Reynolds Equation for pure impact conditions can be written as

$$\frac{\partial}{\partial x}\left(a\frac{\rho h^3}{12\eta}\frac{\partial p}{\partial x}\right)-a\frac{\partial \rho h}{\partial t}=0, \tag{2.1}$$

with the pressure $p(t, x)$, the density $\rho(p)$, the viscosity $\eta(p)$, and the gap height $h(t, x)$, where $t$ denotes the time and $x$ is the spatial coordinate. The factor $a$ is introduced to express a single equation for the line contact problem and for the axisymmetric point contact problem.[1] It is defined as

$$a=\begin{cases}1, & \text{for a line contact, and}\\ x, & \text{for an axisymmetric point contact.}\end{cases} \tag{2.2}$$

The first term of the Reynolds Equation describes the pressure induced flow and is commonly referred to as the Poiseuille flow term. The second term is commonly referred to as the squeeze term.

To solve the above initial-boundary value problem, boundary and initial conditions need to be prescribed. Pressure boundary conditions are prescribed on the boundary enclosing the spatial domain $\Omega_{\mathrm{F}} := [x_{\mathrm{L}},\ x_{\mathrm{R}}]$. Initial conditions are prescribed at the initial time instant $t_0$ for the entire spatial domain $\Omega_{\mathrm{F}}$. For the line contact problem, pressure boundary conditions are applied on both sides of the domain, i.e.,

$$\begin{aligned}p(x_{\mathrm{L}},t)&=p_{\mathrm{L}},\ \forall t\geq t_0,\ \text{and}\\ p(x_{\mathrm{R}},t)&=p_{\mathrm{R}},\ \forall t\geq t_0.\end{aligned} \tag{2.3}$$

For the axisymmetric point contact problem, the pressure gradient is prescribed to vanish at the axis of symmetry. Here, the axis of symmetry is $x_{\mathrm{L}}=0$, i.e.,

$$\left.\frac{\partial p(x,t)}{\partial x}\right|_{x=x_{\mathrm{L}}}=0,\ \forall t\geq t_0. \tag{2.4}$$

The boundary condition at the outer boundary of the domain remains as given in Eq. (2.3). Please note that a vanishing pressure gradient can also be prescribed for the line contact problem if the geometry is symmetric and the pressure boundary conditions are equal.

For the initial time instant $t_0$, the initial pressure distribution $p_0(x)$ is prescribed

$$p(x,t_0)=p_0(x),\ \forall x\in\Omega_{\mathrm{F}}. \tag{2.5}$$

[1] Please note that for the axisymmetric point contact problem the spatial coordinate $x$ is the radial coordinate with origin at the center of the contact, i.e., $x=0$.

### Cavitation condition

The Reynolds Equation is derived under the assumption of a fully-filled fluid gap. Physically, this assumption does not necessarily hold true throughout the whole spatial and temporal domain. Hence, for a temporally and spatially diverging gap, i.e., $\partial h/\partial t > 0$ and $\partial h/\partial x > 0$, the Reynolds Equation may predict negative pressure values. In reality, fluids cannot sustain considerable tension and the fluid would cavitate, preventing the pressure to drop significantly under the cavitation pressure $p_{\text{cav}}$. To prevent the Reynolds Equation from yielding unphysical results, additional constraints need to be introduced.

It is difficult to describe the phenomenon of cavitation precisely. Numerous research has been carried out to gain a better understanding of the related phenomena. In the field of thin film lubrication Swift-Stieber [120, 122] (also called Reynolds conditions) and Jakobsson-Floberg-Olsson (JFO) [78, 102] conditions are most commonly used. Swift-Stieber conditions aim to predict the location of film rupture, whereas JFO conditions also try to predict the location of film reformation. Thus, JFO conditions aim to enforce mass-conservation.

The review paper by Braun [12] gives a more detailed overview of the topic.

In this work, Swift-Stieber conditions are used. These conditions can be written as

$$\begin{aligned} p(x,t) &\geq p_{\text{cav}},\ \forall t \geq t_0 \text{ and } \forall x \in \Omega_{\text{F}}, \\ \left.\frac{\partial p(x,t)}{\partial x}\right|_{x=x_{\text{cav}}} &= 0,\ \forall t \geq t_0, \end{aligned} \tag{2.6}$$

where $x_{\text{cav}}$ denotes the location of film rupture. Due to the high pressures in the contact region, $p_{\text{cav}} = 0$ Pa is usually assumed. These conditions can be enforced by adding a penalty term to the right side of the Reynolds Equation [137]

$$\frac{\partial}{\partial x}\left(a\frac{\rho h^3}{12\eta}\frac{\partial p}{\partial x}\right) - a\frac{\partial \rho h}{\partial t} = \gamma_{\text{pen}} \min\left(p,\, 0\right). \tag{2.7}$$

The penalty factor $\gamma_{\text{pen}}$ should be chosen as large as possible to enforce the condition properly. However, attention needs to be paid as a very large penalty factor increases the numerical stiffness of the problem, which may cause numerical issues.

## 2.1.2 Lubricant properties

The lubricant properties—especially the viscosity—vary significantly with pressure. In the lubricated impact problem, the pressure variation in space and time is significant. Therefore, the pressure induced variation of lubricant properties needs to be considered.

### Viscosity

A widely used viscosity model was presented by Barus [6] in 1893

$$\eta(p) = \eta_0 \exp\left(\alpha_\eta\, p\right), \tag{2.8}$$

with the pressure-viscosity-coefficient $\alpha_\eta$ and viscosity $\eta_0$ at ambient pressure. The exponential approach of Barus can lead to an overestimation of the viscosity for very large pressures. This disadvantage is improved by Roelands' model [111]

$$\eta(p) = \eta_0 \exp\left(\left(\ln\left(\frac{\eta_0}{\text{Pa}\cdot\text{s}}\right) + 9.67\right)\left(-1 + \left(1 + \frac{p}{1.96 \cdot 10^8\ \text{Pa}}\right)^{\beta_\eta}\right)\right), \tag{2.9}$$

with the pressure-viscosity-index

$$\beta_\eta = \frac{\alpha_\eta}{\ln\left(\frac{\eta_0}{\text{Pa}\cdot\text{s}}\right) + 9.67}\, 1.96 \cdot 10^8\ \text{Pa}.$$

Roelands' model is commonly used in EHD contact problems where high pressures occur.

### Density

Even though fluids are typically modeled as incompressible, here, the high pressure compresses the fluid, which leads to an increase in density. This behavior is described by the model of Dowson and Higginson [35]

$$\rho(p) = \rho_0 \frac{5.9 \cdot 10^8\ \text{Pa} + 1.34p}{5.9 \cdot 10^8\ \text{Pa} + p}, \tag{2.10}$$

with density $\rho_0$ at ambient pressure.

## 2.1.3 Nondimensionalization

Very large pressures and very small gap heights are characteristics of the EHD contact problem. The differences in the order of magnitude of the quantities can lead to a poor condition of the numerical problem. To improve the condition number and to reduce the number of parameters, a nondimensionalization is introduced.

The solutions of the highly loaded EHD contact problem are similar to those of the Hertzian contact problem [70]. The characteristic parameters of the Hertzian contact problem are therefore often used for the nondimensionalization.

The nondimensionalization for the lubricated impact problem in this work is based on parameter sets by Wang et al. [129] for the line contact and Venner et al. [125] for the point contact. Both parameter sets are motivated by a dry contact analysis by Johnson [79]. In this context, the Hertzian pressure $p_H$ and the Hertzian half-width $b$, are based on the maximum load $w_{\max}$, which is estimated to occur during the impact-rebound process. The parameters are computed as

$$p_H = \frac{2\, w_{\max}}{\pi\, b}, \quad b = \sqrt{\frac{8\, w_{\max} R_{\text{red}}}{E'\pi}}, \quad \text{for a line contact, and} \tag{2.11}$$

$$p_H = \frac{3\, w_{\max}}{2\pi\, b^2}, \quad b = \left(\frac{3\, w_{\max}}{2\, E'}\right)^{1/3}, \quad \text{for a point contact.} \tag{2.12}$$

Where the maximum load is given as

$$w_{\max} = k\, \delta_{\max}, \quad \text{for a line contact, and} \tag{2.13}$$

$$w_{\max} = k\, \delta_{\max}^{3/2}, \quad \text{for a point contact.} \tag{2.14}$$

The maximum deformation is computed as

$$\delta_{\max} = \sqrt{\frac{m_0}{k}} v_0, \quad \text{with } k = \frac{E'}{1.52}, \quad \text{for a line contact, and} \tag{2.15}$$

$$\delta_{\max} = \left(\frac{5\, m_0 v_0^2}{4\, k}\right)^{2/5}, \quad \text{with } k = \frac{4}{3}\sqrt{R_{\text{red}}} E', \quad \text{for a point contact,} \tag{2.16}$$

with the initial impact velocity $v_0$, the mass $m_0$, the effective Young's modulus $E'$,

$$E' = 2\left(\frac{1-\nu_1^2}{E_1} + \frac{1-\nu_2^2}{E_2}\right)^{-1}, \tag{2.17}$$

and the reduced radius $R_{\text{red}}$,

$$R_{\text{red}} = \left(\frac{1}{R_1} + \frac{1}{R_2}\right)^{-1}. \tag{2.18}$$

With the above relations and the defined parameters, the dimensionless framework is given as

$$x = bX, \quad a = bA, \quad t = \frac{\delta_{\max}}{v_0} T, \quad h = \frac{b^2 H}{R_{\text{red}}}, \quad p = p_H P, \quad \rho = \rho_0 \bar{\rho}, \quad \eta = \eta_0 \bar{\eta}. \tag{2.19}$$

Applying relations (2.19) to Reynolds Equation (2.1) gives the dimensionless Reynolds Equation

$$\frac{\partial}{\partial X}\left(A \frac{\bar{\rho} H^3}{\lambda \bar{\eta}} \frac{\partial P}{\partial X}\right) - A \frac{\partial \bar{\rho} H}{\partial T} - \bar{\gamma}_{\text{pen}} \min(P, 0) = 0, \tag{2.20}$$

with $\lambda = \left(12 v_0 \eta_0 R_{\text{red}}^2\right) / \left(b^2 p_H \delta_{\max}\right)$. In the following, the coefficient $\xi = \left(A\bar{\rho}H^3\right) / (\lambda\bar{\eta})$ is used to improve the readability.

### 2.1.4 Discretization of Reynolds Equation

The Reynolds Equation is approximated spatially at the grid points $X_i$ using finite differences according to [128] and is approximated temporally at the time instances $t_j$ using SDIRK methods according to Section 2.4. The discretized Reynolds Equation at grid point $X_i$ and time instance $t_j$ can be expressed as

$$f_P\Big|_{X_i}^{t_j} = \lambda \frac{\partial}{\partial X}\left(\xi \frac{\partial P}{\partial X}\right)\Bigg|_{X_i}^{t_j} - A\frac{\partial \bar{\rho} H}{\partial T}\Bigg|_{X_i}^{t_j} - \bar{\gamma}_{\text{pen}} \min(P,\, 0)\Big|_{X_i}^{t_j} = 0. \tag{2.21}$$

The Poiseuille term is discretized using second-order accurate central differences

$$\frac{\partial}{\partial X}\left(\xi \frac{\partial P}{\partial X}\right)\Bigg|_{X_i}^{t_j} \approx \frac{\xi_{i-}^{jk} P_{i-1}^{jk} - \left(\xi_{i-}^{jk} + \xi_{i+}^{jk}\right) P_i^{jk} + \xi_{i+}^{jk} P_{i+1}^{jk}}{\Delta X^2}, \tag{2.22}$$

with $\xi_{i\pm}^{jk} = \left(\xi_{i\pm 1}^{jk} + \xi^{jk}{}_i\right)/2$. In the discretized equations the subscript $i$ denotes the approximated value of the quantity at the $i$-th fluid node and, analogously, the superscript $jk$ denotes the approximated value of the quantity at the $k$–th stage of the $j$–th time step.

The temporal derivatives are integrated using SDIRK methods, which are described in Section 2.4.

Boundary conditions are prescribed on the boundary of the dimensionless domain $\partial\bar{\Omega}^F := \left\{X_1,\, X_{n_F}\right\}$ as described in Eqs. (2.3) and (2.4). In the case of symmetry the vanishing pressure gradient at $X = X_1$, i.e.,

$$\frac{\partial P}{\partial X}\Bigg|_{X_1}^{t_j} = 0, \tag{2.23}$$

is enforced by the second-order finite differences approximation of the pressure gradient as

$$f_P\Big|_{X_1}^{t_j} \approx -3P_1^{jk} + 4P_2^{jk} - P_3^{jk} = 0 \tag{2.24}$$

at $X = X_1$ for all stages $k$ of all time steps $j$. For high pressure-viscosity-coefficients, the convergence could be improved by a first-order approximation of the pressure boundary condition.

Ambient pressure in the entire domain is set as the initial condition for the lubricated impact problem.

## 2.2 Structural model

The fluid pressure leads to a deformation of the contacting bodies. In EHD contact problems the contact region is usually small compared to the spatial extension of the bodies. Therefore, the bodies are typically modeled as elastic half-spaces. Modeling the bodies as half-spaces, however, neither considers structural inertia effects nor does it allow for pressure boundary conditions and complex geometries. Additionally, it can be difficult to accurately describe the motion of the rigid body because the bodies' mutual approach is non-trivial to compute [133]. Different from conventional approaches, in this work, the elastic behavior is approximated using the finite element method, which removes the aforementioned drawbacks.

The theory presented in the following is based on the works [7, 68, 76, 142], which provide a more detailed overview.

### 2.2.1 Elasticity relations

Continuum mechanics provides a foundation to describe the dynamics of a deformable body by a macroscopic model. In the following the equations to describe linear elastic material behavior are presented. The problem is defined by kinematic relations, equilibrium equations, and the constitutive model.

In a fixed reference system the position of a material point $\vec{x}(t)$ is given in a Cartesian coordinate system with coordinates $x$, $y$, and $z$ and base vectors $\vec{e_x}$, $\vec{e_y}$, and $\vec{e_z}$ as

$$\vec{x}(t) = x\,(t)\;\vec{e_x} + y\,(t)\;\vec{e_y} + z\,(t)\;\vec{e_z}. \tag{2.25}$$

The displacement of a material point can be described by the displacement vector $\vec{u}\,(t,\,\vec{x})$ as

$$\vec{u}\,(t,\,\vec{x}) = u_x\,(t,\,\vec{x})\;\vec{e_x} + u_y\,(t,\,\vec{x})\;\vec{e_y} + u_z\,(t,\,\vec{x})\;\vec{e_z},$$

with the corresponding displacement components $u_x$, $u_y$, and $u_z$, respectively.

For small deformations, the strains at a given position can be described by the strain tensor

$$\mathbf{E}\,(t,\,\vec{x}) = \frac{1}{2}\left[\operatorname{grad}\vec{u}\,(t,\,\vec{x}) + \operatorname{grad}^{\mathsf{T}}\;\vec{u}\,(t,\,\vec{x})\right], \tag{2.26}$$

with its components

$$\begin{bmatrix} \frac{\partial u_x}{\partial x} & \frac{1}{2}\left(\frac{\partial u_x}{\partial y}+\frac{\partial u_y}{\partial x}\right) & \frac{1}{2}\left(\frac{\partial u_x}{\partial z}+\frac{\partial u_z}{\partial x}\right) \\ \frac{1}{2}\left(\frac{\partial u_y}{\partial x}+\frac{\partial u_x}{\partial y}\right) & \frac{\partial u_y}{\partial y} & \frac{1}{2}\left(\frac{\partial u_y}{\partial z}+\frac{\partial u_z}{\partial y}\right) \\ \frac{1}{2}\left(\frac{\partial u_z}{\partial x}+\frac{\partial u_x}{\partial z}\right) & \frac{1}{2}\left(\frac{\partial u_z}{\partial y}+\frac{\partial u_y}{\partial z}\right) & \frac{\partial u_z}{\partial z} \end{bmatrix} = \begin{bmatrix} \varepsilon_{xx} & \varepsilon_{xy} & \varepsilon_{xz} \\ \varepsilon_{xy} & \varepsilon_{yy} & \varepsilon_{yz} \\ \varepsilon_{xz} & \varepsilon_{yz} & \varepsilon_{zz} \end{bmatrix}. \tag{2.27}$$

The strain tensor is symmetric and thus contains six different strain components, three normal strain components $\varepsilon_{ii}$ and three shear strain components $\varepsilon_{ij}$.

The equilibrium conditions are given as

$$\operatorname{div} \mathbf{T}(t,\ \vec{x}) + \rho(\vec{x})\,\vec{k} = \rho(\vec{x})\,\ddot{\vec{u}}(t,\ \vec{x}), \tag{2.28}$$

with the stress tensor $\mathbf{T}$, the density $\rho$, and the external force per mass $\vec{k}$. Similar to the strain tensor $\mathbf{E}$, the stress tensor $\mathbf{T}$ is symmetric and contains three normal components $\sigma_{ii}$ and three shear components $\tau_{ij}$.

The relations above can describe small deformations of any material. The material model provides a relation between strain and stress.

According to the Voigt notation, displacement, strain, and stress are transformed into column matrix form,

$$\begin{aligned} \boldsymbol{u} &= \left[u_x\ u_y\ u_z\right]^{\mathsf{T}}, \\ \boldsymbol{E} &= \left[\varepsilon_{xx}\ \varepsilon_{yy}\ \varepsilon_{zz}\ 2\varepsilon_{yz}\ 2\varepsilon_{xz}\ 2\varepsilon_{xy}\right]^{\mathsf{T}}, \text{ and} \\ \boldsymbol{T} &= \left[\sigma_{xx}\ \sigma_{yy}\ \sigma_{zz}\ \tau_{yz}\ \tau_{xz}\ \tau_{xy}\right]^{\mathsf{T}}. \end{aligned} \tag{2.29}$$

The differential operator matrix $\mathbf{D} \in \mathbb{R}^{6\times3}$ gives the relation between strain vector $\boldsymbol{E} \in \mathbb{R}^6$ and displacement vector $\boldsymbol{u} \in \mathbb{R}^3$,

$$\boldsymbol{E} = \mathbf{D}\,\boldsymbol{u}, \text{ with } \mathbf{D} = \begin{bmatrix} \frac{\partial}{\partial x} & 0 & 0 \\ 0 & \frac{\partial}{\partial y} & 0 \\ 0 & 0 & \frac{\partial}{\partial z} \\ 0 & \frac{\partial}{\partial z} & \frac{\partial}{\partial y} \\ \frac{\partial}{\partial z} & 0 & \frac{\partial}{\partial x} \\ \frac{\partial}{\partial y} & \frac{\partial}{\partial x} & 0 \end{bmatrix}. \tag{2.30}$$

Using the stress and strain vector, linear elastic material behavior of the continuum can be modeled through

$$\boldsymbol{T} = \mathbf{C}\,\boldsymbol{E}, \tag{2.31}$$

with the symmetric elasticity matrix $\mathbf{C} \in \mathbb{R}^{6\times6}$. The elasticity matrix relates strain and stress in a continuum. Written as $\boldsymbol{E} = \mathbf{C}^{-1}\boldsymbol{T}$, it is commonly referred to as generalized Hooke's law [50, 106]. For isotropic material behavior, the inverse of the elasticity matrix is given as

$$\mathbf{C}^{-1} = \frac{1}{E}\begin{bmatrix} 1 & -\nu & -\nu & 0 & 0 & 0 \\ -\nu & 1 & -\nu & 0 & 0 & 0 \\ -\nu & -\nu & 1 & 0 & 0 & 0 \\ 0 & 0 & 0 & 2(1+\nu) & 0 & 0 \\ 0 & 0 & 0 & 0 & 2(1+\nu) & 0 \\ 0 & 0 & 0 & 0 & 0 & 2(1+\nu) \end{bmatrix}. \tag{2.32}$$

Hence, isotropic linear elastic material behavior is described by only two parameters, namely Young's modulus $E$ and Poisson's ratio $\nu$.

Using the above relations and suitable initial and boundary conditions the behavior of a linear elastically deforming body can be described. The above description assumes a three-dimensional problem. For two-dimensional problems, the relations are slightly different, as shown in [142].

### 2.2.2 Finite element method

For complex geometries, the linear elasticity relations presented above generally cannot be solved analytically. Therefore, the solution is approximated using the finite element method. The following derivation outlines the concept and will help in understanding the concepts of model order reduction and fluid-structure coupling in Subsection 2.3.2 and Section 3.5, respectively.

The starting point of the derivation is the weak form of Eq. (2.28) given by

$$\int_V \delta\vec{u}\cdot\rho\ddot{\vec{u}}\,\mathrm{d}V + \int_V \delta\mathbf{E}\cdot\mathbf{T}\,\mathrm{d}V - \int_V \delta\vec{u}\cdot\rho\vec{k}\,\mathrm{d}V - \int_A \delta\vec{u}\cdot\vec{t}\,\mathrm{d}A = 0, \tag{2.33}$$

with the virtual displacement $\delta\vec{u}$, the virtual strain $\delta\mathbf{E} = \frac{1}{2}(\mathrm{grad}\delta\vec{u} + \mathrm{grad}^T\delta\vec{u})$, surface load density $\vec{t} := \mathbf{T}\,\vec{n}$, and the surface normal vector $\vec{n}$. To approximate this equation on an arbitrarily shaped domain, the domain is approximated by small, simply shaped elements. This means that the three-dimensional domain $V$ is approximated by the discrete domain $\Omega_{\mathrm{S}}$, made up of subdomains $\Omega_e$. Analogously, the surface $A$ is approximated by $\Gamma$. The column matrix $\boldsymbol{x} \in \mathbb{R}^3$ contains the coordinates of a point in the discrete three-dimensional domain.

Shape functions are used to interpolate the displacement within the discrete domain.[2] The shape functions can be organized in a matrix $\mathbf{N}_e(\boldsymbol{x}) \in \mathbb{R}^{3 \times n_{\text{en}}}$. This matrix contains the value of the shape functions evaluated at $\boldsymbol{x} \in \mathbb{R}^3$ for all nodes $n_{\text{en}}$ of an element $e$. The displacements $\boldsymbol{u}^h \in \mathbb{R}^3$ in the discrete domain are given as

$$\boldsymbol{u}^h(t, \boldsymbol{x}) = \mathbf{N}_e(\boldsymbol{x})\, \boldsymbol{u}_e(t), \tag{2.34}$$

where the superscript $h$ indicates the discretized domain. The element displacement vector $\boldsymbol{u}_e \in \mathbb{R}^{3 \cdot n_{\text{en}}}$ contains the three displacement components of all nodes of an element. The approximated displacements $\boldsymbol{u}^h$ are inserted into Eq. (2.30) to approximate the strains

$$\boldsymbol{E}^h(\boldsymbol{x}) = \mathbf{D}\,\mathbf{N}_e(\boldsymbol{x})\, \boldsymbol{u}_e = \mathbf{B}_e(\boldsymbol{x})\, \boldsymbol{u}_e, \tag{2.35}$$

with the strain-displacement matrix $\mathbf{B}_e(\boldsymbol{x}) \in \mathbb{R}^{6 \times 3 \cdot n_{\text{en}}}$. The approximation is also applied to the accelerations $\ddot{\boldsymbol{u}}^h$, the virtual displacements $\delta \boldsymbol{u}^h$ and the virtual strains $\delta \boldsymbol{E}^h$.

Using the approximated variables and the constitutive relation (2.30), the weak form (2.33) can be written for each element $e$ as

$$\begin{aligned} \delta \boldsymbol{u}_e^{\mathsf{T}} \Bigg\{ \int_{\Omega_e} \mathbf{N}^{\mathsf{T}}{}_e(\boldsymbol{x}) \rho \mathbf{N}_e(\boldsymbol{x})\, \mathrm{d}\Omega\, \ddot{\boldsymbol{u}}_e(t) + \int_{\Omega_e} \mathbf{B}^{\mathsf{T}}{}_e(\boldsymbol{x}) \mathbf{C} \mathbf{B}_e(\boldsymbol{x})\, \mathrm{d}\Omega\, \boldsymbol{u}_e(t) \\ - \int_{\Omega_e} \mathbf{N}^{\mathsf{T}}{}_e(\boldsymbol{x}) \rho \boldsymbol{k}\, \mathrm{d}\Omega - \int_{\Gamma_e} \mathbf{N}^{\mathsf{T}}{}_e(\boldsymbol{x}) \boldsymbol{t}(\boldsymbol{x}, t) \mathrm{d}\Gamma \Bigg\} = 0 \end{aligned} \tag{2.36}$$

The virtual displacements are arbitrary, thus, the expression inside the parenthesis must vanish to fulfill Eq. (2.36). The relations for the element mass matrix, the element stiffness matrix, and the element force vector are given as

$$\mathbf{M}_e = \int_{\Omega_e} \mathbf{N}^{\mathsf{T}}{}_e(\boldsymbol{x})\, \rho\, \mathbf{N}_e(\boldsymbol{x})\, \mathrm{d}\Omega, \tag{2.37}$$

$$\mathbf{K}_e = \int_{\Omega_e} \mathbf{B}^{\mathsf{T}}{}_e(\boldsymbol{x})\, \mathbf{C}\, \mathbf{B}_e(\boldsymbol{x})\, \mathrm{d}\Omega, \text{ and} \tag{2.38}$$

$$\boldsymbol{F}_e = \int_{\Omega_e} \mathbf{N}^{\mathsf{T}}{}_e(\boldsymbol{x})\, \rho\, \boldsymbol{k}\, \mathrm{d}\Omega + \int_{\Gamma_e} \mathbf{N}^{\mathsf{T}}{}_e(\boldsymbol{x})\, \boldsymbol{t}(\boldsymbol{x})\, \mathrm{d}\Gamma, \tag{2.39}$$

respectively.

[2]The finite element method and shape functions are relevant for the fluid-structure coupling which is presented in Section 3.5. Hence, Section 3.5 discusses shape functions in further detail. Figure 3.2 shows the shape functions of a quadratic element.

For each element, a coordinate transformation is carried out and the integrals are approximated using Gauss quadrature. The resulting element matrices are merged to a global matrix. Simplified, the assembly process can be illustrated as follows

$$\mathbf{M} = \cdots + \begin{bmatrix} \mathbf{0} & \mathbf{0} & \cdots & \mathbf{0} \\ \mathbf{0} & \mathbf{M}_{e_1} & \mathbf{0} & \vdots \\ \vdots & \mathbf{0} & \iddots & \vdots \\ \mathbf{0} & \cdots & \cdots & \mathbf{0} \end{bmatrix} + \begin{bmatrix} \mathbf{0} & \cdots & \cdots & \mathbf{0} \\ \vdots & \iddots & \mathbf{0} & \vdots \\ \vdots & \mathbf{0} & \mathbf{M}_{e_2} & \mathbf{0} \\ \mathbf{0} & \cdots & \mathbf{0} & \mathbf{0} \end{bmatrix} + \cdots,$$

where $\mathbf{M}_{e_1}$ and $\mathbf{M}_{e_2}$ are element mass matrices of two different elements. The assembly operator $\bigcup$ indicates this process

$$\mathbf{M} = \bigcup_{e=1}^{n_{\mathrm{el}}} \mathbf{M}_e, \quad \mathbf{K} = \bigcup_{e=1}^{n_{\mathrm{el}}} \mathbf{K}_e, \quad \boldsymbol{F} = \bigcup_{e=1}^{n_{\mathrm{el}}} \boldsymbol{F}_e, \tag{2.40}$$

where $n_{\mathrm{el}}$ is the number of elements and $\mathbf{M} \in \mathbb{R}^{n_u \times n_u}$, $\mathbf{K} \in \mathbb{R}^{n_u \times n_u}$, $\boldsymbol{F} \in \mathbb{R}^{n_u}$ are the global mass matrix, global stiffness matrix, and global force vector, respectively. The element displacement vectors $\boldsymbol{u}_e \in \mathbb{R}^{n_{\mathrm{en}}}$ and element acceleration vectors $\ddot{\boldsymbol{u}}_e \in \mathbb{R}^{n_{\mathrm{en}}}$ are also assembled to a global displacement vector $\boldsymbol{u} \in \mathbb{R}^{n_u}$ and a global acceleration vector $\ddot{\boldsymbol{u}} \in \mathbb{R}^{n_u}$, where $n_u$ denotes the displacement degrees of freedom. With the assembled matrices (2.40), the behavior of the discretized body can be described by a set of ordinary differential equations

$$\mathbf{M}\,\ddot{\boldsymbol{u}} + \mathbf{K}\,\boldsymbol{u} = \boldsymbol{F}. \tag{2.41}$$

### 2.2.3 Gap height

In lubricated contacts, the distance between the contacting surfaces is needed for the computation of the fluid pressure. The distance between the surfaces is referred to as the gap height $h(t, x)$. The gap height is computed as

$$h(t, x) = h_{\mathrm{geo}}(x) + \delta(t, x), \tag{2.42}$$

with the undeformed gap height $h_{\mathrm{geo}}$ and the gap displacement $\delta(t,\ x)$. In this work, the line contact problem and the axisymmetric point contact problem are investigated. In both cases, the position is given by the scalar $x$. Figure 2.1 shows exemplarily the two deformed bodies and the resulting gap height.

The gap displacement is computed as

$$\delta(t,\ x) = u_{z,[1]}(t, x) + u_{z,[2]}(t, x), \tag{2.43}$$

with the displacement in the contact region of the lower and upper body, $u_{z,[1]}$ and $u_{z,[2]}$, respectively.

The undeformed gap height is computed as

$$h_{\text{geo}}(x) = z_{\text{geo},[1]}(x) + z_{\text{geo},[2]}(x), \tag{2.44}$$

with the coordinates of the undeformed lower and upper body, $z_{\text{geo},[1]}$ and $z_{\text{geo},[2]}$, respectively.

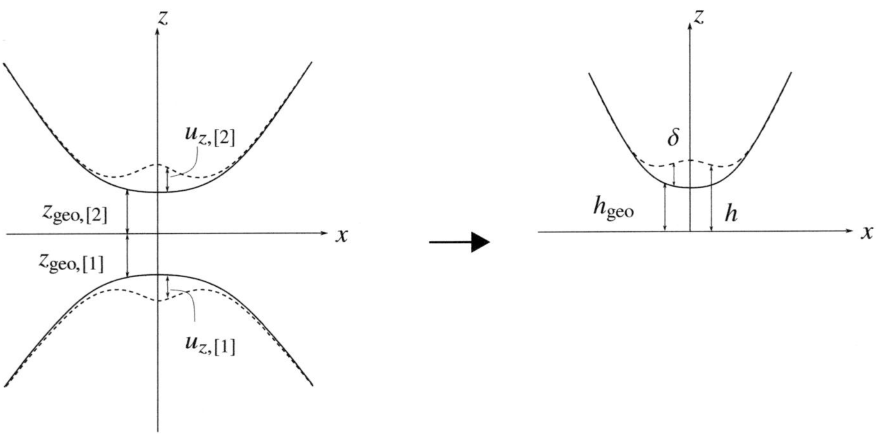


**Figure 2.1:** The displacements of the lower and upper body, i.e., $u_{z,[1]}$ and $u_{z,[2]}$, are combined via Eq. (2.43) to give the gap displacement $\delta$. Analogously, the undeformed geometries of the lower and upper bodies, i.e., $z_{\text{geo},[1]}$ and $z_{\text{geo},[2]}$, are combined to give the undeformed gap height $h_{\text{geo}}$ using Eq. (2.44). The sum of both quantities gives the deformed gap height $h$.

## 2.3 Model order reduction

Model order reduction techniques are generally applied to reduce the computational cost of a model. The idea of most model order reduction techniques is to find a transformation which projects the solution space of the full system onto a lower dimensional space. The lower dimensional space should be as small as possible, while still allowing for a sufficient approximation of the full solution.

The first part of this section outlines the concept of projection onto a lower dimensional subspace. Subsequently, in Subsection 2.3.2, this concept is applied to the structural model. The

Subsections 2.3.3 through 2.3.5 aim to illustrate the construction of a projection matrix for the use in structural dynamics.

The theory presented in the following is based on the works [3, 26, 27, 106], which provide a more detailed overview.

## 2.3.1 Projection

The projection of the full system onto a lower dimensional space is carried out via a projection matrix $\mathbf{V}_z \in \mathbb{R}^{n_z \times \widetilde{n}_z}$. The model defined on the lower dimensional space is called the reduced order model. Its state is given by the reduced state vector $\tilde{z} \in \mathbb{R}^{\widetilde{n}_z}$. The reduced state vector is related to the full state vector $z \in \mathbb{R}^{n_z}$ via

$$z \approx \mathbf{V}_z \, \tilde{z}, \tag{2.45}$$

where $n_z$ is the dimension of the full system and $\widetilde{n}_z$ the dimension of the reduced system. For the projection to be worthwhile, $\widetilde{n}_z \ll n_z$ should hold.

Inserting projection (2.45) into the full system

$$\boldsymbol{f}(z) = \mathbf{0}, \tag{2.46}$$

with the differentiable function $\boldsymbol{f} \in \mathbb{R}^{n_z}$ yields

$$\boldsymbol{f}(\mathbf{V}_z \, \tilde{z}) = \boldsymbol{r}, \tag{2.47}$$

where $\boldsymbol{r}$ is the residuum resulting from the approximation. An additional constraint is needed to obtain a unique solution of the over-determined system of equations (2.47). To this end, a so-called test space $\mathbf{W}_z \in \mathbb{R}^{n_z \times \widetilde{n}_z}$ is introduced. The test-space is required to be orthogonal to the residuum

$$\mathbf{W}_z^\mathsf{T} \, \boldsymbol{r} \overset{!}{=} \mathbf{0}. \tag{2.48}$$

In this work, $\mathbf{W}_z = \mathbf{V}_z$ is chosen, which is referred to as a Galerkin projection. Thus, the reduced order model can be expressed as

$$\mathbf{V}_z^\mathsf{T} \, \boldsymbol{f}(\mathbf{V}_z \, \tilde{z}) = \mathbf{0}. \tag{2.49}$$

## 2.3.2 Application to a structural model

The discretization of complex geometries often requires many elements. This leads to a high number of degrees of freedom, which have to be solved for in every time step of a transient analysis. For specific applications, the number of degrees of freedom can be reduced using model order reduction procedures. For lubricated contacts, the behavior of the bodies in the contact area is of interest. Figure 2.2 illustrates a reduced order model which expresses the solid's behavior at the degrees of freedom in the contact region.

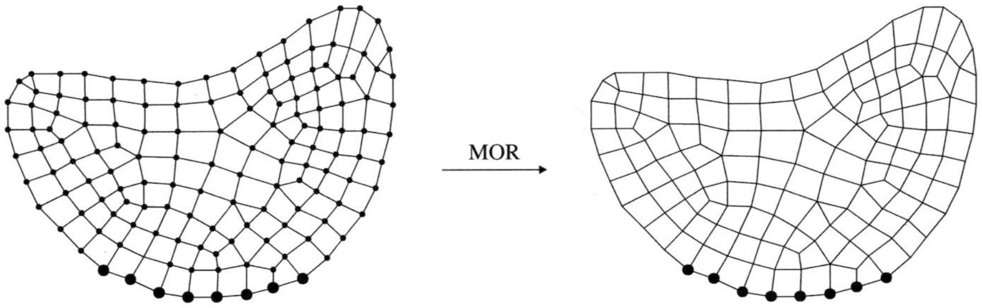


**Figure 2.2:** Model order reduction (MOR) techniques aim to express the behavior of a high-dimensional model (left) by a reduced order model (right). The reduced order model shall consist of a much lower number of degrees of freedom. This is exemplarily shown for a structural finite element model, where only the behavior of the highlighted nodes is of interest. Hence, the MOR procedure aims to express the behavior in this region with as few degrees of freedom as possible.

The body's displacement is approximated by the equation of motion of the undamped system (2.41)

$$\mathbf{M}\,\ddot{\boldsymbol{u}} + \mathbf{K}\,\boldsymbol{u} = \boldsymbol{F}. \tag{2.50}$$

As in Eq. (2.45) a projection is applied, i.e.,

$$\boldsymbol{u} \approx \mathbf{V}_u\,\tilde{\boldsymbol{u}}. \tag{2.51}$$

Via the projection matrix $\mathbf{V}_u \in \mathbb{R}^{n_u \times \widetilde{n}_u}$, the displacements of the reduced order model $\tilde{\boldsymbol{u}} \in \mathbb{R}^{\widetilde{n}_u}$ approximate the displacements of the full model $\boldsymbol{u} \in \mathbb{R}^{n_u}$. Using a Galerkin projection, the full system (2.50) is projected onto the reduced space as

$$\mathbf{V}_u^\mathsf{T}\,\mathbf{M}\,\mathbf{V}_u\,\ddot{\tilde{\boldsymbol{u}}} + \mathbf{V}_u^\mathsf{T}\,\mathbf{K}\,\mathbf{V}_u\,\tilde{\boldsymbol{u}} = \mathbf{V}_u^\mathsf{T}\,\boldsymbol{F}, \tag{2.52}$$

where $\widetilde{n}_u$ gives the size of the reduced space. Using the substitution in Eq. (2.52), the reduced order model can be expressed as

$$\tilde{\mathbf{M}}\,\ddot{\tilde{\boldsymbol{u}}} + \tilde{\mathbf{K}}\,\tilde{\boldsymbol{u}} = \tilde{\boldsymbol{F}}, \tag{2.53}$$

with the reduced mass matrix $\tilde{\mathbf{M}} \in \mathbb{R}^{\widetilde{n}_u \times \widetilde{n}_u}$, reduced stiffness matrix $\tilde{\mathbf{K}} \in \mathbb{R}^{\widetilde{n}_u \times \widetilde{n}_u}$, and reduced force vector $\tilde{\boldsymbol{F}} \in \mathbb{R}^{\widetilde{n}_u}$.

Now, a suitable projection matrix for the structural model has to be constructed. The following shows different approaches for constructing a projection matrix for structural dynamics problems.

### 2.3.3 Modal truncation

Discretization using FEM yields a set of coupled differential equations, see Eq. (2.50). Solving the eigenvalue problem

$$\left(\mathbf{K} - \omega_i^2 \mathbf{M}\right) \boldsymbol{\varphi}_i = \mathbf{0}, \text{ with } i = 1, 2, 3, \ldots, n_{\mathrm{nm}}, \tag{2.54}$$

yields the natural eigenfrequency $\omega_i$ and the corresponding eigenvector $\boldsymbol{\varphi}_i \in \mathbb{R}^{n_u}$ for $n_{\mathrm{nm}}$ modes. The model's transient behavior can be approximated using modal superposition. To this end, only a small number of modes $n_{\mathrm{nm}}$ is kept and the rest is neglected, where usually $n_{\mathrm{nm}} \ll n_u$ holds. This procedure is generally known as modal truncation. The modes are normalized with respect to the mass matrix and collected in the normal mode matrix

$$\boldsymbol{\Phi}_{\mathrm{nm}} = \begin{bmatrix} \boldsymbol{\varphi}_1 & \boldsymbol{\varphi}_2 & \ldots & \boldsymbol{\varphi}_{n_{\mathrm{nm}}} \end{bmatrix}. \tag{2.55}$$

The resulting modes are called *normal modes* and have the properties

$$\boldsymbol{\varphi}^{\mathsf{T}}{}_i \, \mathbf{M} \, \boldsymbol{\varphi}_j = \delta_{ij}, \text{ and} \tag{2.56}$$

$$\boldsymbol{\varphi}^{\mathsf{T}}{}_i \, \mathbf{K} \, \boldsymbol{\varphi}_j = \omega_i^2 \delta_{ij}, \tag{2.57}$$

where $\delta_{ij}$ is the Kronecker delta. In modal truncation, the normal mode matrix $\boldsymbol{\Phi}_{\mathrm{nm}} \in \mathbb{R}^{n_u \times n_{\mathrm{nm}}}$ is used as projection matrix $\mathbf{V}_u = \boldsymbol{\Phi}_{\mathrm{nm}}$. Applying the projection matrix to the equation of motion (2.50) yields the equation of motion in modal space

$$\ddot{\tilde{\boldsymbol{u}}} + \boldsymbol{\Lambda} \, \tilde{\boldsymbol{u}} = \mathbf{V}^{\mathsf{T}}{}_u \, \boldsymbol{F}, \tag{2.58}$$

with the spectral matrix $\boldsymbol{\Lambda} = \operatorname{diag}\left(\omega_1^2, \omega_2^2, \ldots, \omega_{n_{\mathrm{nm}}}^2\right) \in \mathbb{R}^{n_{\mathrm{nm}} \times n_{\mathrm{nm}}}$. The spectral matrix is a diagonal matrix containing the squared eigenfrequencies as diagonal elements. The equation of motion in modal space is uncoupled. This property can speed up the computation significantly. However, the reduced order model is only valid in the close-range of the state around which it was constructed. The limited validity makes the approach rather unsuitable for problems where the boundary conditions change significantly, e.g., contact problems.

For more information on mode shapes and modal truncation, see [44] and the literature cited therein.

### 2.3.4 Static condensation and Guyan reduction

Static condensation is helpful to understand an essential idea in model order reduction of structural mechanics. It is the concept of separating the model's degrees of freedom into master and slave degrees of freedom. This method can be useful when only small regions of the body are of interest, as exemplarily shown in Fig. 2.2. The nodes in the region of interest are referred to as master nodes. Their displacements are summarized in the vector $\boldsymbol{u}_{\mathrm{m}} \in \mathbb{R}^{n_{\mathrm{m}}}$, with the number of master degrees of freedom $n_{\mathrm{m}}$. The goal is to obtain the displacements of the master nodes without having to solve all degrees of freedom of the complete model.

Partitioning the static equations of equilibrium into master (subscript m) and slave (subscript s) degrees of freedom gives

$$\begin{bmatrix} \mathbf{K}_{\mathrm{mm}} & \mathbf{K}_{\mathrm{ms}} \\ \mathbf{K}_{\mathrm{sm}} & \mathbf{K}_{\mathrm{ss}} \end{bmatrix} \begin{Bmatrix} \boldsymbol{u}_{\mathrm{m}} \\ \boldsymbol{u}_{\mathrm{s}} \end{Bmatrix} = \begin{Bmatrix} \boldsymbol{F}_{\mathrm{m}} \\ \boldsymbol{F}_{\mathrm{s}} \end{Bmatrix}. \tag{2.59}$$

The slave degrees of freedom are to be expressed in terms of master degrees of freedom. Matrix operations can achieve this objective. When there are no forces at the slave degrees of freedom, i.e., $\boldsymbol{F}_{\mathrm{s}} = \mathbf{0}$, the following relation holds

$$\boldsymbol{u}_{\mathrm{s}} = -\mathbf{K}_{\mathrm{ss}}^{-1} \mathbf{K}_{\mathrm{sm}} \boldsymbol{u}_{\mathrm{m}}. \tag{2.60}$$

Using this relation, the constraint mode matrix is created as

$$\boldsymbol{\Phi}_{\mathrm{cm}} = \begin{bmatrix} \mathbf{I} \\ -\mathbf{K}_{\mathrm{ss}}^{-1} \mathbf{K}_{\mathrm{sm}} \end{bmatrix}. \tag{2.61}$$

The columns of matrix $\boldsymbol{\Phi}_{\mathrm{cm}} \in \mathbb{R}^{n_u \times n_{\mathrm{m}}}$ are commonly referred to as *constraint modes*, see [27].[3] In static condensation, the constraint mode matrix is used as the projection matrix, i.e., $\mathbf{V}_u = \boldsymbol{\Phi}_{\mathrm{cm}}$. This MOR technique is referred to as static condensation. Since only matrix operations on the stiffness matrix are carried out, the static condensation is exact for static problems.

Natural mode analysis of large finite element models can take up a significant amount of time. Therefore, it is also desirable to reduce the degrees of freedom for natural mode analysis. With this motivation, Guyan [54] used the same projection matrix as in static condensation to reduce the mass matrix. Applying the constraint mode matrix to the mass matrix yields

$$\tilde{\mathbf{M}} = \begin{bmatrix} \mathbf{I} & -\mathbf{K}_{\mathrm{sm}}^{\mathsf{T}} \mathbf{K}_{\mathrm{ss}}^{-\mathsf{T}} \end{bmatrix} \begin{bmatrix} \mathbf{M}_{\mathrm{mm}} & \mathbf{M}_{\mathrm{ms}} \\ \mathbf{M}_{\mathrm{sm}} & \mathbf{M}_{\mathrm{ss}} \end{bmatrix} \begin{bmatrix} \mathbf{I} \\ -\mathbf{K}_{\mathrm{ss}}^{-1} \mathbf{K}_{\mathrm{sm}} \end{bmatrix}, \tag{2.62}$$

with the reduced mass matrix $\tilde{\mathbf{M}} \in \mathbb{R}^{\tilde{n}_u \times \tilde{n}_u}$. For this reduction technique, $\tilde{n}_u = n_{\mathrm{m}}$ holds.

However, the constraint mode matrix (2.61) is derived from static equilibrium (2.59). Applying the same transformation to the dynamic equilibrium is inconsistent and yields a reduced mass

[3] Constraint modes can also be constructed by prescribing a unit displacement to one master degree of freedom and fixing the other master degrees of freedom. The resulting displacement vector of the full model would be equivalent to one column of $\boldsymbol{\Phi}_{\mathrm{cm}}$.

matrix with mixed stiffness and mass components, see Eq. (2.62). As a consequence, this approach may yield erroneous results for dynamic analyses. This limitation was also observed by Guyan in his original paper [54] and other authors after that, e.g., [19].

While this technique can be helpful in some problems, the dynamic properties of the reduced model are strongly dependent on the number and location of the chosen master degrees of freedom as well as the stiffness and mass distribution of the body, see [106]. Generally, this approach yields acceptable results in the low frequency range but is not suitable for higher frequencies. The approximation quality with respect to natural mode analysis increases as more, spatially distributed, master degrees of freedom are chosen.

### 2.3.5 Component mode synthesis

As in the Guyan reduction, the concept of component mode synthesis (CMS) is to approximate the behavior of the slave degrees of freedom with as few unknowns as possible. Component mode synthesis combines the beneficial properties of the aforementioned methods: Modal truncation and Guyan reduction.

The Guyan reduction uses constraint modes to approximate the dynamic behavior. Constraint modes may yield bad results for dynamic problems. However, the static displacement is approximated well by constraint modes.

Modal truncation uses normal modes, which approximate the unconstrained and not externally excited dynamic behavior well. However, they yield bad results for problems with changing boundary conditions, e.g., contact problems.

Component mode synthesis aims to improve upon the drawbacks of both methods by combining different types of modes. Here, normal modes $\mathbf{\Phi}_{\mathrm{nm}}$ are used to approximate the dynamics of the structure and constraint modes $\mathbf{\Phi}_{\mathrm{cm}}$ are introduced to better approximate the static and overall behavior. In CMS, the projection matrix is generally given as

$$\mathbf{V}_{\mathrm{CMS}} = \begin{bmatrix} \mathbf{\Phi}_{\mathrm{nm}} & \mathbf{\Phi}_{\mathrm{cm}} \end{bmatrix}. \tag{2.63}$$

Component mode synthesis methods can be classified according to the boundary conditions prescribed to the master degrees of freedom—also referred to as interface nodes—during normal mode extraction. They are classified into fixed-interface methods, free-interface methods or hybrid methods, see [25]. The various options to create and combine modes result in a variety of CMS methods. The Craig-Bampton method [4] is a popular CMS method. It is a simplification of Hurty's method [77] and is based on fixed-interface modes. Also widely used are the free-interface methods by Craig-Chang [25], MacNeal [92] and Rubin [113]. The review papers [26, 28] give an overview of the most common approaches.

In this work, two methods are used: the Craig-Bampton method and an adapted Craig-Chang method.[4] Both methods utilize constraint modes and normal modes. The Craig-Bampton method uses fixed-interface normal modes. The fixed-interface normal modes are created by constraining the master degrees of freedom during the eigenvalue computation. Therefore, they result from the eigenvalue problem of the slave structure, i.e.,

$$(\mathbf{K}_{\text{ss}} - \omega_i^2 \mathbf{M}_{\text{ss}})\boldsymbol{\varphi}_i = \mathbf{0}, \text{ with } i = 1, 2, \ldots, n_{\text{nm}}, \tag{2.64}$$

where $n_{\text{nm}}$ denotes the number of normal modes. After normalization with respect to mass according to Eq. (2.56), the kept modes are assembled in the fixed-interface normal mode matrix

$$\boldsymbol{\Phi}_{\text{fixed}} = \begin{bmatrix} & \mathbf{0} & & \\ \boldsymbol{\varphi}_1 & \boldsymbol{\varphi}_2 & \ldots & \boldsymbol{\varphi}_{n_{\text{nm}}} \end{bmatrix}. \tag{2.65}$$

The resulting projection matrix for the Craig-Bampton method is

$$\mathbf{V}_{\text{fixed}} = \begin{bmatrix} \boldsymbol{\Phi}_{\text{fixed}} & \boldsymbol{\Phi}_{\text{cm}} \end{bmatrix}, \tag{2.66}$$

with the constraint mode matrix $\boldsymbol{\Phi}_{\text{cm}} \in \mathbb{R}^{n_u \times n_{\text{cm}}}$ according to Eq. (2.61), plus rigid body modes, if existing. The number of constraint modes $n_{\text{cm}}$ is equal to the number of master degrees of freedom plus rigid body degrees of freedom. The projection matrix is of size $\mathbf{V}_{\text{fixed}} \in \mathbb{R}^{n_u \times \widetilde{n}_u}$ with $\widetilde{n}_u = n_{\text{cm}} + n_{\text{nm}}$.

The adapted Craig-Chang method uses free-interface normal modes. The free-interface normal modes are obtained by solving the eigenvalue problem of the full structure

$$(\mathbf{K} - \omega_i^2 \mathbf{M})\boldsymbol{\varphi}_i = \mathbf{0}, \text{ for } i = 1, 2, \ldots, n_{\text{nm}}. \tag{2.67}$$

However, since the master degrees of freedom are unconstrained during mode extraction, their values may be nonzero, which makes the displacement computation more difficult. Additionally, the free-interface modes and the constraint modes might be linearly dependent. Therefore, the free-interface normal modes are transformed according to [9, 83] into the "fixed-interface form"

$$\boldsymbol{\Phi}_{\text{free}} = \begin{bmatrix} \boldsymbol{\Phi}^{\text{s}}_{\text{free}}(1 : n_{\text{cm}}, :) - \boldsymbol{\Phi}_{\text{cm}}(1 : n_{\text{cm}}, :)\, \boldsymbol{\Phi}^{\text{s}}_{\text{free}}(1 : n_{\text{cm}}, :) \\ \boldsymbol{\Phi}^{\text{s}}_{\text{free}}(n_{\text{cm}} + 1 : n_u, :) - \boldsymbol{\Phi}_{\text{cm}}(n_{\text{cm}} + 1 : n_u, :)\, \boldsymbol{\Phi}^{\text{s}}_{\text{free}}(1 : n_{\text{cm}}, :) \end{bmatrix}.$$

[4]The adapted Craig-Chang method is a free-interface method, which uses constraint modes, see [9, 30, 31]. It is an adapted version of the Craig-Chang method [25]. The adapted Craig-Chang method is used in the commercial finite element software Abaqus. The solution procedure in this work aims to be compatible with Abaqus.

Due to the structure of the constraint mode matrix $\mathbf{\Phi}_{\mathrm{cm}}$, see Eq. (2.61), the modified free-interface normal modes become

$$\mathbf{\Phi}_{\mathrm{free}} = \begin{bmatrix} \mathbf{0} \\ \mathbf{\Phi}^{\mathrm{s}}_{\mathrm{free}}(n_{\mathrm{cm}}+1 : n_u, :) - \mathbf{\Phi}_{\mathrm{cm}}(n_{\mathrm{cm}}+1 : n_u, :)\, \mathbf{\Phi}^{\mathrm{s}}_{\mathrm{free}}(1 : n_{\mathrm{cm}}, :) \end{bmatrix},$$

where $\mathbf{\Phi}^{\mathrm{s}}_{\mathrm{free}} \in \mathbb{R}^{n_u \times n_{\mathrm{nm}}}$ contains $n_{\mathrm{nm}}$ unchanged free-interface normal modes. This operation does not change the subspace spanned by the projection matrix [9, 83].

After orthogonalization and normalization, the modified free-interface normal modes are included in the projection matrix for the Craig-Chang method. The projection matrix is given as

$$\mathbf{V}_{\mathrm{free}} = \begin{bmatrix} \mathbf{\Phi}_{\mathrm{free}} & \mathbf{\Phi}_{\mathrm{cm}} \end{bmatrix}, \tag{2.68}$$

where $\mathbf{V}_{\mathrm{free}} \in \mathbb{R}^{n_u \times \widetilde{n}_u}$ with $\widetilde{n}_u = n_{\mathrm{cm}} + n_{\mathrm{nm}}$.

The main difference between the two methods arise from the state of the boundary conditions of the master nodes during modal analysis, i.e., fixed-interface normal modes versus free-interface normal modes.

## 2.4 Time integration

Spatial discretization of the fluid and structural equations yields a system of ordinary differential equations.[5] In the following, this system equations is integrated in time. This work proposes the use of singly diagonal implicit Runge-Kutta methods (SDIRK) for time integration. This class of methods is well suited for stiff problems, such as the coupled problem at hand. Additionally, they allow for higher-order time integration and for an efficient adaptive time stepping scheme based on embedded methods.

The first part of this section outlines the solution procedure of an initial value problem using SDIRK methods. The second part of this section focuses on the adaptive time step size control. Table 2.1 summarizes the steps of the procedure and helps to understand different variables.

The theory presented in the following is based on the works [1, 62, 63, 67] , which provide a more detailed overview.

[5] Depending on the set of equations, spatial discretization may yield a system of differential-algebraic equations. The implications of this case are studied in [39, 40, 67].

**Table 2.1:** Overview of the implementation of the time integration scheme.

Initialize initial values ($\boldsymbol{y}_0$, $t_0$) and time integration method, see Tab. 2.3

Loop over time steps: $j = j + 1$ until $t(j) = t_{\text{end}}$

- Loop over stages: $k = 1, \ldots, s$
  - Compute starting vector, see Eq. (2.72) $\rightsquigarrow \boldsymbol{S}^{jk}$
    $\boldsymbol{S}^{jk} = \boldsymbol{y}_j + \Delta t_j \sum_{l=1}^{k-1} a_{kl} \dot{\boldsymbol{Y}}^{jl}$
  - Solve system of equations $\rightsquigarrow \boldsymbol{Y}^{jl}$
    see Section 3.6
  - Compute stage derivatives, see Eq. (2.73) $\rightsquigarrow \dot{\boldsymbol{Y}}^{jl}$
    $\dot{\boldsymbol{Y}}^{jl} = \frac{\boldsymbol{Y}^{jl} - \boldsymbol{S}^{jk}}{\Delta t_j \, a_{kk}},$
  - Compute stage value, see Eq. (2.74) $\rightsquigarrow \boldsymbol{Y}^{jk}$
    $\boldsymbol{Y}^{jk} = \boldsymbol{S}^{jk} + \Delta t_j \, a_{kk} \, \dot{\boldsymbol{Y}}^{jk}$
- Compute next time step, see Eq. (2.70) $\rightsquigarrow \boldsymbol{y}_{j+1}$
  $\boldsymbol{y}_{j+1} = \boldsymbol{y}_j + \Delta t_j \sum_{k=1}^{s} b_k \, \dot{\boldsymbol{Y}}^{jk}$
- Step size control
  - Estimate local integration error, see Eq. (2.77) $\rightsquigarrow \hat{\boldsymbol{y}}_j^{\text{err}}$
    $\hat{\boldsymbol{y}}_j^{\text{err}} = \Delta t_j \sum_{k=1}^{s} \left(\hat{b}_k - b_k\right) \dot{\boldsymbol{Y}}^{jk}$
  - Compute time step size for next time step $\rightsquigarrow \Delta t_{j+1}$
    Two approaches are given via Eqs. (2.79) and (2.80)
  - Accept solution? $\rightsquigarrow t_{j+1} = t_j + \Delta t_j$
    According to the estimated local integration error the solution may be rejected and recomputed with a smaller time step size.

### 2.4.1 Singly diagonal implicit Runge-Kutta method

The initial value problem

$$\dot{\boldsymbol{y}} = \boldsymbol{f}(t, \boldsymbol{y}(t)), \quad \boldsymbol{y}(t_0) = \boldsymbol{y}_0, \tag{2.69}$$

with initial conditions $\boldsymbol{y}_0 \in \mathbb{R}^{n_y}$ and $t_0$ is solved in the time interval $t_0 \leq t \leq t_{\text{end}}$. Within the interval, the solution of (2.69) is approximated at time instances $t_j$. The integer $j = 0, 1, \ldots, n_t$ indicates the current time step. Starting from the initial conditions $\boldsymbol{y}_0 = \boldsymbol{y}(t_0)$, the solution is approximated at the next time instance $t_{j+1}$ by

$$\boldsymbol{y}\left(t_{j+1}\right) \approx \boldsymbol{y}_{j+1} = \boldsymbol{y}_j + \Delta t_j \sum_{k=1}^{s} b_k \boldsymbol{f}\left(T_{jk}, \boldsymbol{Y}^{jk}\right), \tag{2.70}$$

with the time step size $\Delta t_j = t_{j+1} - t_j$, the number of stages $s$, the weighting factors $b_k$, the stage time $T_{jk} = t_j + c_k \Delta t_j$ and the stage values $\boldsymbol{Y}^{jk} \in \mathbb{R}^{n_y}$.

The stage values are computed as

$$\boldsymbol{Y}^{jk} = \boldsymbol{y}_j + \Delta t_j \sum_{l=1}^{k} a_{kl} \dot{\boldsymbol{Y}}^{jl}, \tag{2.71}$$

with the weights $a_{kl}$ for each stage $k = 1, 2, \ldots, s$. The auxiliary variable $l = 1, 2, \ldots, k$ counts up until the current stage. In diagonally implicit Runge-Kutta methods the entries in the Butcher tableau above the diagonal are zero, i.e., $a_{kl} = 0$ if $l > k$. This helpful property allows the computation of stage values one after the other. In this context, it is useful to introduce the starting vector $\boldsymbol{S}^{jk} \in \mathbb{R}^{n_y}$

$$\boldsymbol{S}^{jk} = \boldsymbol{y}_j + \Delta t_j \sum_{l=1}^{k-1} a_{kl} \dot{\boldsymbol{Y}}^{jl}, \tag{2.72}$$

which is made up of the solutions of the previous stages. Thus, the starting vector is known at the beginning of each stage value computation - analogous to the initial values.

The stage derivatives $\dot{\boldsymbol{Y}}^{jl} \in \mathbb{R}^{n_y}$ can be expressed as

$$\dot{\boldsymbol{Y}}^{jl} = \frac{\boldsymbol{Y}^{jl} - \boldsymbol{S}^{jk}}{\Delta t_j \, a_{kk}}. \tag{2.73}$$

Generally, the stage derivatives $\dot{\boldsymbol{Y}}^{jl}$ are substituted into the nonlinear system of equations and then the system is solved for the stage values $\boldsymbol{Y}^{jl}$, see Section 3.6. From the stage values the stage derivatives can be computed using Eq. (2.73). The stage derivatives can also be computed from the initial value problem. However, attention needs to be paid to evaluate the problem at the proper stage time and stage value, i.e., $\dot{\boldsymbol{Y}}^{jk} = \boldsymbol{f}(T_{jk}, \boldsymbol{Y}^{jk})$.

Combining Eq. (2.72) and Eq. (2.71) yields the stage value

$$\boldsymbol{Y}^{jk} = \boldsymbol{S}^{jk} + \Delta t_j \, a_{kk} \, \dot{\boldsymbol{Y}}^{jk}, \tag{2.74}$$

In SDIRK methods, the following relations hold $a_{kl} = 0$ if $l > k$ and $a_{sl} = b_l$, see Tab. 2.2. Therefore, the stage values at the final stage $s$ are identical to the approximation of the solution at the next time step, i.e., $\boldsymbol{y}_{j+1} = \boldsymbol{Y}^{js}$.

The methods are defined by weighting factors $a_{kl}$, $b_k$, and $c_k$. The weights are given in a Butcher tableau, see Tab. 2.2. Table 2.3 gives the Butcher tableaus of the methods used in this work.

**Table 2.2:** The weights defining a Runge-Kutta method are summarized in a Butcher tableau.

$$\begin{array}{c|cccc} c_1 & a_{11} & a_{12} & \cdots & a_{1s} \\ c_2 & a_{21} & a_{22} & \cdots & a_{2s} \\ \vdots & \vdots & \vdots & \ddots & \vdots \\ c_s & a_{s1} & a_{s2} & \cdots & a_{ss} \\ \hline & b_1 & b_2 & \cdots & b_s \\ & \hat{b}_1 & \hat{b}_2 & \cdots & \hat{b}_s \end{array}$$

**(a)** Butcher tableau of a general Runge–Kutta method.

$$\begin{array}{c|cccc} c_1 & a_{11} & 0 & \cdots & 0 \\ c_2 & a_{21} & a_{22} & \ddots & \vdots \\ \vdots & \vdots & & \ddots & 0 \\ c_s & b_1 & b_2 & \cdots & b_s \\ \hline & b_1 & b_2 & \cdots & b_s \\ & \hat{b}_1 & \hat{b}_2 & \cdots & \hat{b}_s \end{array}$$

**(b)** Butcher tableau of an SDIRK method.

### 2.4.2 Adaptive step size selection

The choice of the time step size $\Delta t_j$ influences the quality of the approximation of the solution and the computational cost of the problem. A too large time step size might yield unusable results, whereas a too small time step size is computationally inefficient. The choice of a suitable constant time step size becomes even more difficult when the velocity of the dynamic processes changes during the solution process. Hence, it is desirable to use an algorithm that adjusts the time step size automatically for a given problem. Such algorithms depend on an indicator on the basis of which the size of the time step can be selected. The local integration error of a time step can be used as an indicator. However, the cost of computing the indicator should be low.

For SDIRK methods embedded schemes can be derived. Embedded schemes allow for an approximation of the local integration error at hardly any additional cost. Based on the methods of Alexander [1], Ellsiepen [39] proposed an embedded first-order method. The Butcher tableau of this method is given in Tab. 2.3.

The local integration error gives the difference between the exact solution and the numerical solution, i.e.,

$$\boldsymbol{y}^{\text{err}}(t_j) = \boldsymbol{y}(t_j) - \boldsymbol{y}_j. \tag{2.75}$$

With the solution of an embedded method $\hat{\boldsymbol{y}}_j \in \mathbb{R}^{n_y}$, an error estimator of the form

$$\hat{\boldsymbol{y}}_j^{\text{err}} = \hat{\boldsymbol{y}}_j - \boldsymbol{y}_j \tag{2.76}$$

can be easily computed, where $\boldsymbol{y}_j \in \mathbb{R}^{n_y}$ and $\hat{\boldsymbol{y}}_j \in \mathbb{R}^{n_y}$ are approximated solutions, obtained with an SDIRK method of order $p$ and the corresponding embedded method of order $\hat{p}$, respectively, with $\hat{p} < p$. The embedded methods are designed such that the weights $a_{kl}$ and $c_k$ are the same as in the full order method, only the weights $b_k$ differ. Therefore, the error estimator (2.76) can be computed at hardly any extra cost as

$$\hat{\boldsymbol{y}}_j^{\text{err}} = \Delta t_j \sum_{k=1}^{s} \left(\hat{b}_k - b_k\right) \boldsymbol{f}\left(T^{jk}, \boldsymbol{Y}^{jk}\right). \tag{2.77}$$

### Time step size control

An error measure $err_y$ is computed as

$$err_y = \text{rms}\left(\frac{\hat{\boldsymbol{y}}_j^{\text{err}}}{\epsilon_r \left\|\hat{\boldsymbol{y}}_j\right\| + \epsilon_a}\right). \tag{2.78}$$

Here, the error measure is based on the root mean square, which is computed via Eq. (A.1). The error measure relates the approximated local integration error $\hat{\boldsymbol{y}}_j^{\text{err}}$ to the relative tolerance $\epsilon_{\text{r}}$ and the absolute tolerance $\epsilon_{\text{a}}$. In multi-field problems, the order of magnitude of the field variables may differ. In such cases, it can be helpful to compute an error measure for each quantity [32]. In this work, error measures are computed for pressure, $err_p$, displacement, $err_u$, and generalized coordinates, $err_q$. The maximum of these error measures $err_m = \max(err_p,\ err_u,\ err_q)$ is used to find a new time step size.

One approach of obtaining a new time step size is given by

$$\Delta t_{j+1} = \Delta t_j \begin{cases} f_{\min} & \text{for } err_m > 1, \\ 1 & \text{for } 0.1 \leq err_m \leq 1, \\ f_{\max} & \text{for } err_m < 0.1. \end{cases} \tag{2.79}$$

In the range $0.1 \leq err_m \leq 1$, the time step size is kept constant, and for larger or smaller values, it is reduced or increased by the factor $f_{\min}$ or $f_{\max}$, respectively. In this approach, the factors by which the time step size is adjusted are constant throughout the computation and independent of the order of the time integration scheme.

Another approach is to adapt the time step size such that the local integration error stays as close as possible to the prescribed tolerances. An adaptation of this so-called "standard time step size controller", see [62, p.124], is given by [67] as

$$\Delta t_{j+1} = \Delta t_j \begin{cases} \max\left(f_{\text{min}},\ f_{\text{safety}}\, err_m^{-1/(\hat{p}+1)}\right) & \text{for } err_m > 1, \\ \min\left(f_{\text{max}},\ f_{\text{safety}}\, err_m^{-1/(\hat{p}+1)}\right) & \text{for } err_m \leq 1. \end{cases} \tag{2.80}$$

The parameters $f_{\text{min}}$, $f_{\text{max}}$, and $f_{\text{safety}}$ are prescribed to avoid oscillations or large changes of the time step size. If $err_m > 1$, the time step is rejected and recomputed. Else, the time step is accepted and the next time step computed. In both cases, the time step size $\Delta t_{j+1}$ is used, which is computed using Eq. (2.80). Further time step size control approaches are discussed in [116].

**Table 2.3:** Butcher tableaus of the SDIRK methods used in this work. The methods are of order $p$ and have $s$ stages. The embedded methods are of order $\hat{p}$.

Alexander's second-order method with Ellsiepen's embedded scheme [1, 39]
($p = 2,\ \hat{p} = 1,\ s = 2$)

$$\begin{array}{c|cc} \alpha & \alpha & \\ 1 & 1-\alpha & \alpha \\ \hline & 1-\alpha & \alpha \\ & 1-\hat{\alpha} & \hat{\alpha} \end{array} \qquad \begin{aligned} \alpha &= 1-\sqrt{2}/2 \\ \hat{\alpha} &= 2-5\sqrt{2}/2 \end{aligned}$$

Hairer and Wanner's forth-order method [62, p. 100]
($p = 4,\ \hat{p} = 3,\ s = 5$)

$$\begin{array}{c|ccccc} 1/4 & 1/4 & & & & \\ 3/4 & 1/2 & 1/4 & & & \\ 11/20 & 15/20 & -1/25 & 1/4 & & \\ 1/2 & 371/1360 & -137/2720 & 15/544 & 1/4 & \\ 1 & 25/24 & -49/48 & 125/16 & -85/12 & 1/4 \\ \hline & 25/24 & -49/48 & 125/16 & -85/12 & 1/4 \\ & 59/48 & -17/96 & 225/32 & -85/12 & 0 \end{array}$$

# 3 Solution procedure

The objective of this work is to develop an efficient numerical solution procedure for the lubricated impact problem incorporating the dynamic behavior of complex elastic bodies. This chapter presents the developed solution procedure. Sections 3.1 through 3.3 explain the build-up of the extended structural model, which considers structural inertia effects. The resulting structural model shall be coupled with the fluid model in order to solve the EHD contact problem. To this end, a coupling procedure is developed. Section 3.4 describes a preparation step to increase the efficiency of the coupled problem. Subsequently, the coupling procedure is explained in Section 3.5. The monolithic coupling approach yields a fully coupled problem, which is solved in Section 3.6. Finally, Section 3.7 introduces a step to further reduce the computational time via model order reduction of the complete nonlinear system of equations.

## 3.1 Set-up of the reduced structural model

The proposed workflow starts with creating geometric bodies in a CAE software and subsequently discretizing them using finite elements. According to Subsection 2.2.2, this process yields a global mass matrix $\mathbf{M} \in \mathbb{R}^{n_u \times n_u}$ and a global stiffness matrix $\mathbf{K} \in \mathbb{R}^{n_u \times n_u}$, where $n_u$ gives the displacement degrees of freedom. Based on the forces $\boldsymbol{F} \in \mathbb{R}^{n_u}$, the unknown displacements $\boldsymbol{u} \in \mathbb{R}^{n_u}$ can be computed via the equation of motion of the discretized body

$$\mathbf{M}\ddot{\boldsymbol{u}} + \mathbf{K}\boldsymbol{u} = \boldsymbol{F}. \tag{3.1}$$

In the case of lubricated contacts, the displacements in the contact region are of interest. The discretized body introduces many degrees of freedom outside the contact region. In an attempt to minimize the number of degrees of freedom of the problem, a reduced order model of the finite element model is constructed. Section 2.3 gives an overview of reduction procedures for structural bodies used within this work. In this step, component mode synthesis is applied to construct the reduced order model, see Subsection 2.3.5. The reduced order model is defined by the reduced mass matrix $\tilde{\mathbf{M}} \in \mathbb{R}^{\tilde{n}_u \times \tilde{n}_u}$, the reduced stiffness matrix $\tilde{\mathbf{K}} \in \mathbb{R}^{\tilde{n}_u \times \tilde{n}_u}$, and the reduced force vector $\tilde{\boldsymbol{F}} \in \mathbb{R}^{\tilde{n}_u}$, where $\tilde{n}_u$ gives the number of the reduced degrees of freedom.

Using the reduced matrices, the behavior of the structural body can be approximated by the reduced order model

$$\tilde{\mathbf{M}}\ddot{\tilde{\boldsymbol{u}}} + \tilde{\mathbf{K}}\tilde{\boldsymbol{u}} = \tilde{\boldsymbol{F}}, \tag{3.2}$$

with the reduced displacements $\tilde{\boldsymbol{u}} \in \mathbb{R}^{\widetilde{n}_u}$.

## 3.2 Uncoupled equations of motion

In the previous step, the body's geometry was created and a reduced order model was constructed. In this step, the resulting system of equations (3.2) is uncoupled via transformation into the modal space. This change of basis is effectively the same procedure as described in Subsection 2.3.3 on modal truncation. However, different from modal truncation, all modes are kept. Solving the eigenvalue problem (2.54) with the reduced mass and stiffness matrix, $\tilde{\mathbf{M}}$ and $\tilde{\mathbf{K}}$, yields the modal matrix

$$\bar{\boldsymbol{\Phi}} = \begin{bmatrix} \hat{\varphi}_1 & \hat{\varphi}_2 & \dots & \hat{\varphi}_{n_q} \end{bmatrix}, \tag{3.3}$$

which consists of the mass normalized eigenvectors $\hat{\boldsymbol{\varphi}}_i \in \mathbb{R}^{n_q}$, ordered corresponding to their eigenvalue $\omega_i$ in an ascending manner. The modal matrix $\bar{\boldsymbol{\Phi}} \in \mathbb{R}^{n_q \times n_q}$ is used as the projection matrix according to Eq. (2.52). It transforms the equation of motion (3.2) into the modal space. The reduced mass matrix and the reduced stiffness matrix are diagonalized by the system's eigenvectors and become

$$\bar{\boldsymbol{\Phi}}^{\mathsf{T}}\tilde{\mathbf{M}}\bar{\boldsymbol{\Phi}} = \mathbf{I} \text{ and } \bar{\boldsymbol{\Phi}}^{\mathsf{T}}\tilde{\mathbf{K}}\bar{\boldsymbol{\Phi}} = \boldsymbol{\Lambda}. \tag{3.4}$$

The reduced stiffness matrix in modal space $\overline{\mathbf{K}} \in \mathbb{R}^{n_q \times n_q}$ is the spectral matrix $\boldsymbol{\Lambda} = \mathrm{diag}\left(\omega_1^2, \omega_2^2, \dots, \omega_{n_q}^2\right) \in \mathbb{R}^{n_q \times n_q}$, containing the squared eigenvalues of the system as diagonal elements, and the reduced mass matrix in modal space $\overline{\mathbf{M}} \in \mathbb{R}^{n_q \times n_q}$ is the identity matrix.

Consequently, the resulting equations of motion in modal space can be written as

$$\begin{aligned} \ddot{\boldsymbol{q}} + \overline{\mathbf{K}}\,\boldsymbol{q} &= \bar{\boldsymbol{\Phi}}^{\mathsf{T}}\,\tilde{\boldsymbol{F}}, \\ \tilde{\boldsymbol{u}} &= \bar{\boldsymbol{\Phi}}\,\boldsymbol{q}, \end{aligned} \tag{3.5}$$

with the generalized coordinates $\boldsymbol{q} \in \mathbb{R}^{n_q}$. No modes are truncated in this step. Hence, the transformation of the equations of motion into the modal space does not reduce the number of degrees of freedom, i.e., $\widetilde{n}_u = n_q$. Within the framework of this work, the transformation into the modal space primarily brings the following advantages:

- The uncoupled representation can significantly reduce the computational effort to solve the system of equations. This reduction becomes especially relevant when the matrices are not sparse, which is the case for the reduced order model at hand. It needs to be considered that a modal analysis must be carried out, which may require significant computational time. However, the modal analysis only needs to be carried out once in the solution procedure.
- Resulting from the constraint modes, artificial frequencies are introduced into the model, which need to be treated as covered in Section 3.3, to improve the solution quality. The uncoupling of the equations is a prerequisite for this step.

Besides these, there are more advantages and disadvantages of mode-based linear dynamic analyses, which are further discussed in [7, 44].

## 3.3 Reducing spurious oscillations

The previous section concluded with a reduced order model in modal space. Especially in highly dynamic problems, reduced order models containing constraint modes can show spurious high-frequency oscillations, see Subsection 4.2.3. These spurious modes are numerical artifacts that are introduced into the model by the constraint modes, i.e., the dynamic contribution of the constraint modes may yield spurious oscillations. However, the static contribution of the constraint modes is needed to approximate the local deformation. Therefore, it is desirable to consider their static contribution while limiting their dynamic contribution. This section outlines approaches to minimize the dynamic contribution of the constraint modes in order to reduce spurious oscillations.

### Separating the modes

The modal matrix $\bar{\mathbf{\Phi}} \in \mathbb{R}^{n_q \times n_q}$ is divided into two parts. These parts will be referred to as dynamically considered modes $\bar{\mathbf{\Phi}}_{\mathrm{dc}} \in \mathbb{R}^{n_q \times n_{\mathrm{dc}}}$ and statically considered modes $\bar{\mathbf{\Phi}}_{\mathrm{sc}} \in \mathbb{R}^{n_q \times n_{\mathrm{sc}}}$, with $\bar{\mathbf{\Phi}} := \begin{bmatrix} \bar{\mathbf{\Phi}}_{\mathrm{dc}} & \bar{\mathbf{\Phi}}_{\mathrm{sc}} \end{bmatrix}$. The division of the modes has a decisive influence on the quality of the solution. If too many modes are considered dynamically, unwanted oscillations occur, if too few are considered, the system loses part of its dynamic properties. Here, $n_{\mathrm{dc}}$ gives the number of dynamically considered modes. The dynamically considered modes $\bar{\mathbf{\Phi}}_{\mathrm{dc}}$ are defined as the first $n_{\mathrm{dc}}$ columns of the modal matrix, $\bar{\mathbf{\Phi}}_{\mathrm{dc}} := \left[\bar{\mathbf{\Phi}}\left(:,\ 1 : n_{\mathrm{dc}}\right)\right]$. Consequently, the statically considered modes are defined as the remaining columns of the modal matrix $\bar{\mathbf{\Phi}}_{\mathrm{sc}} := \left[\bar{\mathbf{\Phi}}\left(:,\ n_{\mathrm{dc}} + 1 : n_q\right)\right]$. The number of statically considered modes is given as $n_{\mathrm{sc}} = n_q - n_{\mathrm{dc}}$.

Accordingly, the generalized coordinates are divided into two parts

$$\boldsymbol{q} = \left\{ \begin{array}{c} \boldsymbol{q}_{\mathrm{dc}} \\ \boldsymbol{q}_{\mathrm{sc}} \end{array} \right\},$$

where $\boldsymbol{q}_{\mathrm{dc}} \in \mathbb{R}^{n_{\mathrm{dc}}}$ are the generalized coordinates of the dynamically considered modes and $\boldsymbol{q}_{\mathrm{sc}} \in \mathbb{R}^{n_{\mathrm{sc}}}$ corresponds to the statically considered modes. Applying the same procedure to Eq. (3.5) and introducing the modal damping matrix $\overline{\mathbf{D}}$ yields

$$\left\{ \begin{array}{c} \ddot{\boldsymbol{q}}_{\mathrm{dc}} \\ \ddot{\boldsymbol{q}}_{\mathrm{sc}} \end{array} \right\} + \left[ \begin{array}{cc} \overline{\mathbf{D}}_{\mathrm{dc}} & \mathbf{0} \\ \mathbf{0} & \overline{\mathbf{D}}_{\mathrm{sc}} \end{array} \right] \left\{ \begin{array}{c} \dot{\boldsymbol{q}}_{\mathrm{dc}} \\ \dot{\boldsymbol{q}}_{\mathrm{sc}} \end{array} \right\} + \left[ \begin{array}{cc} \overline{\mathbf{K}}_{\mathrm{dc}} & \mathbf{0} \\ \mathbf{0} & \overline{\mathbf{K}}_{\mathrm{sc}} \end{array} \right] \left\{ \begin{array}{c} \boldsymbol{q}_{\mathrm{dc}} \\ \boldsymbol{q}_{\mathrm{sc}} \end{array} \right\} = \bar{\mathbf{\Phi}}^{\mathsf{T}} \tilde{\boldsymbol{F}}, \tag{3.6a}$$

$$\tilde{\boldsymbol{u}} = \bar{\mathbf{\Phi}} \left\{ \begin{array}{c} \boldsymbol{q}_{\mathrm{dc}} \\ \boldsymbol{q}_{\mathrm{sc}} \end{array} \right\}. \tag{3.6b}$$

The modal damping matrix is constructed as

$$\overline{\mathbf{D}} = \operatorname{diag}(2\zeta_i\omega_i) \text{ with } i = 1,\ 2,\ \ldots,\ n_q, \tag{3.7}$$

with the damping ratio $\zeta_i$ and the eigenfrequency $\omega_i$ of the $i$-th mode. Modal damping allows damping each mode independently. This property is helpful for the next step.

**Treating the different parts of the modal matrix**

In the following, two approaches to minimize the spurious oscillations are outlined:

- The *damping approach* aims to reduce the oscillations by introducing damping into the system. This procedure has been applied to a single node contact problem with one artificial frequency in [84].
- The *feedthrough approach* takes advantage of the division of the modal matrix. This approach only considers the statically considered modes during time integration. It was motivated by [140, 141].

*The damping approach*

An intuitive approach in dealing with the spurious oscillations is to introduce damping into the model. In the damping approach, modal damping, see Eq. (3.7), with a large damping ratio $\zeta_{\mathrm{sc}}$ is prescribed to the statically considered modes $\bar{\mathbf{\Phi}}_{\mathrm{sc}}$. The large damping ratio significantly limits the dynamic influence of the statically considered modes.

*The feedthrough approach*

In the feedthrough approach, dynamic effects resulting from the statically considered modes $\bar{\mathbf{\Phi}}_{\mathrm{sc}}$ are considered as negligible, i.e., $\ddot{\boldsymbol{q}}_{\mathrm{sc}} = \dot{\boldsymbol{q}}_{\mathrm{sc}} = \mathbf{0}$. This assumption transforms Eq. (3.6b) to

$$\ddot{\boldsymbol{q}}_{\mathrm{dc}} + \overline{\mathbf{D}}_{\mathrm{dc}}\,\dot{\boldsymbol{q}}_{\mathrm{dc}} + \overline{\mathbf{K}}_{\mathrm{dc}}\,\boldsymbol{q}_{\mathrm{dc}} = \bar{\mathbf{\Phi}}_{\mathrm{dc}}^{\mathsf{T}}\,\tilde{\boldsymbol{F}}, \tag{3.8}$$

$$\overline{\mathbf{K}}_{\mathrm{sc}}\,\boldsymbol{q}_{\mathrm{sc}} = \bar{\mathbf{\Phi}}_{\mathrm{sc}}^{\mathsf{T}}\,\tilde{\boldsymbol{F}}, \tag{3.9}$$

$$\tilde{\boldsymbol{u}} = \bar{\mathbf{\Phi}} \begin{Bmatrix} \boldsymbol{q}_{\mathrm{dc}} \\ \boldsymbol{q}_{\mathrm{sc}} \end{Bmatrix}, \tag{3.10}$$

which can be written as

$$\ddot{\boldsymbol{q}}_{\mathrm{dc}} + \overline{\mathbf{D}}_{\mathrm{dc}}\,\dot{\boldsymbol{q}}_{\mathrm{dc}} + \overline{\mathbf{K}}_{\mathrm{dc}}\,\boldsymbol{q}_{\mathrm{dc}} = \bar{\mathbf{\Phi}}_{\mathrm{dc}}^{\mathsf{T}}\,\tilde{\boldsymbol{F}}, \tag{3.11a}$$

$$\tilde{\boldsymbol{u}} = \bar{\mathbf{\Phi}}_{\mathrm{dc}}\,\boldsymbol{q}_{\mathrm{dc}} + \bar{\mathbf{\Phi}}_{\mathrm{sc}}\,\overline{\mathbf{K}}_{\mathrm{sc}}^{-1}\,\bar{\mathbf{\Phi}}_{\mathrm{sc}}^{\mathsf{T}}\,\tilde{\boldsymbol{F}}. \tag{3.11b}$$

This formulation can reduce the computational time considerably. The reduced computational effort results from the smaller size of system (3.11a).

### Transforming the system into the state-space representation

The resulting structural model is given as a system of ordinary second-order differential equations, see Eq. (3.6) or Eq. (3.11). In the next step, the system is transformed into the state-space representation. The transformation into the state-space representation allows the application of general solution procedures, such as the time integration schemes presented in Section 2.4. Additionally, it will prove to be useful for the efficient solution procedure of the coupled problem, see Section 3.4.

Substituting the generalized coordinates as

$$\boldsymbol{y}_1 = \boldsymbol{q}, \tag{3.12}$$

$$\boldsymbol{y}_2 = \dot{\boldsymbol{y}}_1 = \dot{\boldsymbol{q}}, \tag{3.13}$$

a state-space representation can be written as

$$\begin{bmatrix} \dot{\boldsymbol{y}}_1 \\ \dot{\boldsymbol{y}}_2 \end{bmatrix} = \begin{bmatrix} \mathbf{A}_{11} & \mathbf{A}_{12} \\ \mathbf{A}_{21} & \mathbf{A}_{22} \end{bmatrix} \begin{bmatrix} \boldsymbol{y}_1 \\ \boldsymbol{y}_2 \end{bmatrix} + \begin{bmatrix} \mathbf{B}_1 \\ \mathbf{B}_2 \end{bmatrix} \tilde{\boldsymbol{F}}, \tag{3.14a}$$

$$\tilde{\boldsymbol{u}} = \mathbf{C}\,\boldsymbol{y}_1 + \mathbf{D}\,\tilde{\boldsymbol{F}}, \tag{3.14b}$$

with the state vector

$$\boldsymbol{y} = \begin{Bmatrix} \boldsymbol{y}_1 \\ \boldsymbol{y}_2 \end{Bmatrix} \in \mathbb{R}^{n_y}.$$

and matrices $\mathbf{A} \in \mathbb{R}^{n_y \times n_y}$, $\mathbf{B} \in \mathbb{R}^{n_y \times n_q}$, $\mathbf{C} \in \mathbb{R}^{n_q \times n_y/2}$, and $\mathbf{D} \in \mathbb{R}^{n_q \times n_q}$, where $n_y$ denotes the dimensions of the state-space representation. Note that, the matrices depend on the treatment approach.

For the **damping approach**, the matrices are given as

$$\begin{aligned} \mathbf{A} &= \begin{bmatrix} \mathbf{0} & \mathbf{I} \\ -\bar{\mathbf{K}} & -\bar{\mathbf{D}} \end{bmatrix}, \text{ with } \mathbf{A} \in \mathbb{R}^{n_y \times n_y}, \\ \mathbf{B} &= \begin{bmatrix} \mathbf{0} \\ \bar{\mathbf{\Phi}}^{\mathsf{T}} \end{bmatrix}, \text{ with } \mathbf{B} \in \mathbb{R}^{n_y \times n_{\mathrm{cm}}}, \end{aligned} \tag{3.15}$$

and

$$\tilde{\boldsymbol{u}} = \bar{\mathbf{\Phi}}\, \boldsymbol{y}_1, \tag{3.16}$$

with the dimensions of the state-space representation $n_y = 2 \cdot (n_{\mathrm{dc}} + n_{\mathrm{sc}})$.

For the **feedthrough approach**, the matrices become

$$\begin{aligned} \mathbf{A} &= \begin{bmatrix} \mathbf{0} & \mathbf{I} \\ -\bar{\mathbf{K}}_{\mathrm{dc}} & -\bar{\mathbf{D}}_{\mathrm{dc}} \end{bmatrix}, \text{ with } \mathbf{A} \in \mathbb{R}^{n_y \times n_y}, \\ \mathbf{B} &= \begin{bmatrix} \mathbf{0} \\ \bar{\mathbf{\Phi}}_{\mathrm{dc}}^{\mathsf{T}} \end{bmatrix}, \text{ with } \mathbf{B} \in \mathbb{R}^{n_y \times n_{\mathrm{cm}}}, \end{aligned} \tag{3.17}$$

and

$$\tilde{\boldsymbol{u}} = \bar{\mathbf{\Phi}}_{\mathrm{sc}}\, \boldsymbol{y}_1 + \bar{\mathbf{\Phi}}_{\mathrm{sc}}\, \mathbf{K}_{\mathrm{sc}}^{-1}\, \bar{\mathbf{\Phi}}_{\mathrm{sc}}^{\mathsf{T}}\, \tilde{\boldsymbol{F}}, \tag{3.18}$$

with the dimension of the state-space representation $n_y = 2 \cdot n_{\mathrm{dc}}$.

The above approaches illustrate two options to cope with the spurious oscillations. Their characteristics are further studied in Section 4.2.

## 3.4 Time integration of the structural model

The structural model is coupled into the EHD contact model using a monolithic coupling approach. In other words, all equations are combined in a fully coupled system of equations. The straightforward embedding of the reduced order models of both bodies into the system of equations would result in a large system of equations. The solution of a large system of equations generally requires high numerical costs. For this reason, this section makes use of the linearity of the structural model to combine the structural models of both bodies and thereby keep the size of the entire system low.[1]

The previous section concluded with a reduced order model in state-space representation. The formulation as a first-order differential equation allows for the application of SDIRK methods, see Section 2.4. The stage value $\boldsymbol{Y}^{jk} \in \mathbb{R}^{n_y}$ is given by

$$\boldsymbol{Y}^{jk} = \boldsymbol{S}^{jk} + \Delta t_j\, a_{ii} \left( \mathbf{A}\, \boldsymbol{Y}^{jk} + \mathbf{B}\, \tilde{\boldsymbol{F}}^{jk} \right), \tag{3.19}$$

with the starting vector $\boldsymbol{S}^{jk} \in \mathbb{R}^{n_y}$ and the force vector $\tilde{\boldsymbol{F}}^{jk} = \tilde{\boldsymbol{F}}\left(T_{jk}\right)$ at time $T_{jk}$.[2] The indices refer to stage $k$ of time step $j$.

The matrices $\mathbf{A}$ and $\mathbf{B}$ can be split up into components as

$$\mathbf{A} = \begin{bmatrix} \mathbf{A}_{11} & \mathbf{A}_{12} \\ \mathbf{A}_{21} & \mathbf{A}_{22} \end{bmatrix}, \quad \mathbf{B} = \begin{bmatrix} \mathbf{B}_1 \\ \mathbf{B}_2 \end{bmatrix}, \tag{3.20}$$

where $\mathbf{A}_{12} = \mathbf{I}, \mathbf{A}_{11} = \mathbf{0}$, and $\mathbf{B}_1 = \mathbf{0}$ hold for all treatment approaches. Inserting these components into Eq. (3.19) gives

$$-\frac{1}{\Delta t_j\, a_{ii}} \boldsymbol{Y}_1^{jk} + \boldsymbol{Y}_2^{jk} = -\frac{1}{\Delta t_j\, a_{ii}} \boldsymbol{S}_1^{jk}, \tag{3.21}$$

$$\mathbf{A}_{21} \boldsymbol{Y}_1^{jk} + \left( \mathbf{A}_{22} - \frac{1}{\Delta t_j\, a_{ii}} \mathbf{I} \right) \boldsymbol{Y}_2^{jk} = -\frac{1}{\Delta t_j\, a_{ii}} \boldsymbol{S}_2^{jk} - \mathbf{B}_2\, \tilde{\boldsymbol{F}}^{jk}. \tag{3.22}$$

Combining (3.21) and (3.22) and solving for $\boldsymbol{Y}_1^{jk}$ yields

$$\begin{aligned} \boldsymbol{Y}_1^{jk} = {} & \left( \mathbf{A}_{21} + \frac{1}{\Delta t_j\, a_{ii}} \left( \mathbf{A}_{22} - \frac{1}{\Delta t_j\, a_{ii}} \mathbf{I} \right) \right)^{-1} \\ & \left( \frac{1}{\Delta t_j\, a_{ii}} \left( \mathbf{A}_{22} - \frac{1}{\Delta t_j\, a_{ii}} \mathbf{I} \right) \boldsymbol{S}_1^{jk} - \frac{1}{\Delta t_j\, a_{ii}} \boldsymbol{S}_2^{jk} - \mathbf{B}_2\, \tilde{\boldsymbol{F}}^{jk} \right), \end{aligned} \tag{3.23}$$

in which $\boldsymbol{Y}_2^{jk}$ can be computed as

$$\boldsymbol{Y}_2^{jk} = \frac{1}{\Delta t_j\, a_{ii}} \left( \boldsymbol{Y}_1^{jk} - \boldsymbol{S}_1^{jk} \right). \tag{3.24}$$

[1] Parts of this approach have been presented in a similar form in the student work [S3].

[2] To avoid order reductions of the time integration scheme, nonlinear boundary conditions should be integrated, see [112].

From relations (3.23) and (3.24), the state vector $\boldsymbol{Y}^{jk}$ is known. With the state vector $\boldsymbol{Y}^{jk}$, the displacements can be computed using Eq. (3.14b), which is given in discretized form as

$$\tilde{\boldsymbol{u}}^{jk} = \mathbf{C}\,\boldsymbol{Y}_1^{jk} + \mathbf{D}\,\tilde{\boldsymbol{F}}^{jk}. \tag{3.25}$$

Expressions (3.23) through (3.25) give the final form of the structural model, which is later inserted into the coupled problem, see Section 3.6. This formulation of the structural model expresses the displacements as a function of the contact force. After interpolation, the displacements of both bodies can be summed up to give the gap displacements. Finally, only the gap displacements are included in the coupled system. This approach will keep the overall system size low.

Note that, although the displacements can be expressed explicitly from the forces, the entire system is implicit. The reason for this is that the contact forces result from the integration of the fluid pressure and these, in turn, depend on the displacements.

The following sections discuss the procedure for the coupling of the fluid model with the structural model.

## 3.5 Coupling procedure

Previous sections discussed the construction of the extended structural model. The developed structural model shall be incorporated into the lubricated contact problem. To this end, the structural model must be coupled with the fluid model, i.e., the field variables must be known on the structure as well as on the fluid domain. A suitable coupling procedure needs to meet specific demands since the EHD contact problem is very sensitive to changes in the gap height $h$ and the structural model requires a force vector, which depends on the elements used in the finite element method. A coupling method that meets these requirements is developed and described in this section.

First, a computational grid is defined, then the construction of coupling matrices is presented.

### 3.5.1 Definition of the computational grid

The structural model and the fluid model are discretized on different grids. There may exist up to three different grids: the fluid grid, the structure grid of the lower body and that of the upper body. The fluid grid is defined by the grid ratio $i_{\mathrm{GR}}$ and the smallest structure element in the contact region as

$$\Delta x^{\mathrm{F}} = i_{\mathrm{GR}} \min\left(\Delta x^{\mathrm{S}}_{[1],\min}, \Delta x^{\mathrm{S}}_{[2],\min}\right), \tag{3.26}$$

where $\Delta x^{\mathrm{S}}_{[1],\mathrm{min}}$ and $\Delta x^{\mathrm{S}}_{[2],\mathrm{min}}$ are the smallest elements of the lower and upper structure grid, respectively, and $\Delta x^{\mathrm{F}}$ is the length of a fluid element. The fluid grid points are given by $x_{i_\mathrm{F}} = x_\mathrm{L} + (i_\mathrm{F} - 1)\,\Delta x^{\mathrm{F}}$ with $i_\mathrm{F} = 1, 2, \ldots, n_\mathrm{F}$. They are summarized in the vector $\boldsymbol{x}^{\mathrm{F}} \in \mathbb{R}^{n_\mathrm{F}}$. The grid points of the structure grid are summarized in vector $\boldsymbol{x}^{\mathrm{S}} \in \mathbb{R}^{n_\mathrm{S}}$, where $n_\mathrm{S}$ denotes the number of structure nodes.[3]

In order to solve the coupled problem, a grid, on which all variables are known, must be defined. This grid will be called the computational grid. The fluid grid is equidistant and finer than the structure grids. Hence, it is used as the computational grid. In the following, interpolation matrices are constructed, which interpolate the variables from the respective structure grid onto the computational grid, i.e., the fluid grid.

### 3.5.2 Construction of coupling matrices

The coupling procedure should allow an efficient and accurate solution of the coupled problem. To this end, the following aspects need to be taken into account:

- The Reynolds Equation is discretized using second-order finite differences, see Subsection 2.1.4. Furthermore, quadratic shape functions will be used in the context of the finite element discretization. Hence, the coupling procedure should maintain the spatial convergence order of the discretization schemes.
- The additional computational effort introduced by the coupling procedure should be low. Hence, the coupling procedure should not need to be recomputed in each time step or iteration.
- The EHD contact problem is very sensitive to variations in the gap height profile. Consequently, an inaccurate coupling quickly destabilizes the solution procedure. Hence, the coupling should not introduce artificial variations in the gap height profile.

The proposed procedure is inspired by the interpolation and integration approaches of the finite element method, i.e., the use of shape functions for the interpolation of quantities and the use of the Gauss quadrature for the integration of the pressure. Matrices are constructed to interpolate and integrate the quantities between the different grids. Figure 3.1 shows the different grids and indicates how they are connected via the interpolation and integration matrices. The interpolation matrix $\mathbf{G}_{\mathrm{SF}}$ interpolates the displacements from the structure grid to the fluid grid. The pressure integration matrix $\mathbf{G}_{\mathrm{int}}$ integrates the pressure from the fluid grid points to nodal forces at the structure grid points. The build-up of the needed matrices is summarized in the following.[4]

[3] In the following, indices denoting lower and upper body, as well as the superscript $jk$, indicating the current stage of the time integration, are omitted for ease of notation. All coupling matrices are created for both bodies.

[4] Parts of this section have been presented in a similar form in [S6].

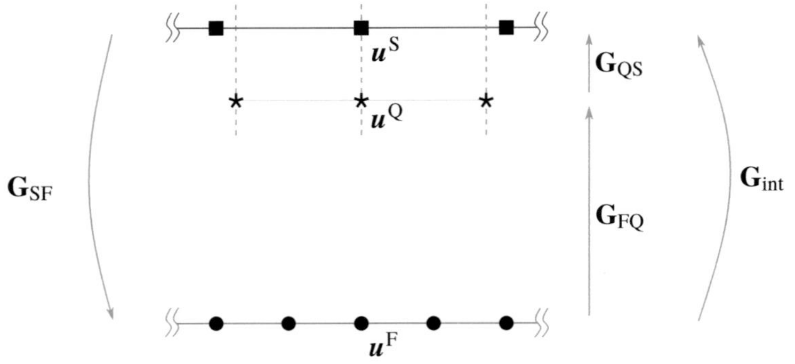


**Figure 3.1:** Schematic sketch of fluid grid (lower), structure grid (upper) and quadrature points on structure grid (middle). The interpolation- and integration-matrices connect the different grids.

### Interpolation matrix $\mathbf{G}_{\mathrm{SF}}$

The geometry and displacement are given on the structure grid $\boldsymbol{x}^{\mathrm{S}} \in \mathbb{R}^{n_{\mathrm{S}}}$, whereas the pressure is computed on the fluid grid $\boldsymbol{x}^{\mathrm{F}} \in \mathbb{R}^{n_{\mathrm{F}}}$. Local shape functions are used for the interpolation of the quantities between grid points. Shape functions for one element can be built from the Lagrange-polynomial

$$N_r(x) = \prod_{\substack{s=1 \\ s \neq r}}^{n_{\mathrm{en}}} \frac{x - x_s}{x_r - x_s}, \tag{3.27}$$

with the nodal coordinates $x_r$ and $x_s$ of node $r$ and $s$ and $n_{\mathrm{en}}$ gives the number of nodes per element.

The shape functions of a quadratic three node element are given by

$$\begin{aligned} N_1^e(x) &= \frac{x - x_2^e}{x_1^e - x_2^e} \frac{x - x_3^e}{x_1^e - x_3^e}, \text{ if } x \in [x_1^e, x_2^e] \text{ else } N_1^e(x) = 0, \\ N_2^e(x) &= \frac{x - x_1^e}{x_2^e - x_1^e} \frac{x - x_3^e}{x_2^e - x_3^e}, \text{ if } x \in [x_1^e, x_2^e] \text{ else } N_2^e(x) = 0, \text{ and} \\ N_3^e(x) &= \frac{x - x_1^e}{x_3^e - x_1^e} \frac{x - x_2^e}{x_3^e - x_2^e}, \text{ if } x \in [x_1^e, x_2^e] \text{ else } N_3^e(x) = 0, \end{aligned} \tag{3.28}$$

with the coordinates $x_1^e$, $x_2^e$, and $x_3^e$ of the left, right, and middle node, respectively.

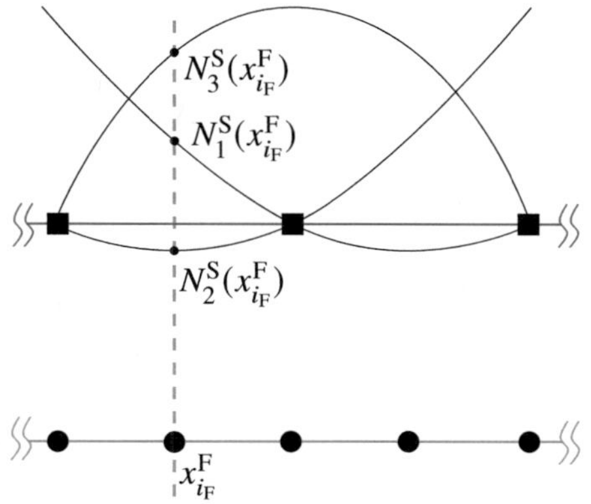


**Figure 3.2:** Quadratic shape functions of a one-dimensional structure element, which consists of three structure nodes. The shape functions are evaluated at the fluid node $x_{i_F}^F$. The values of the shape functions at the fluid node give the weights for the interpolation matrix $\mathbf{G}_{SF}$, see Eq. (3.29).

Figure 3.2 exemplifies the evaluation of the shape functions of a structure element for one fluid node. This evaluation is carried out for each fluid node $x_{i_F}^F$ with $i_F = 1, 2, \ldots, n_F$, using the resulting values the interpolation matrix $\mathbf{G}_{SF} \in \mathbb{R}^{n_F \times n_S}$ is constructed as

$$\mathbf{G}_{SF} = \begin{bmatrix} N_1^S(x_1^F) & N_2^S(x_1^F) & \cdots & N_{n_S}^S(x_1^F) \\ N_1^S(x_2^F) & \vdots & \cdots & N_{n_S}^S(x_2^F) \\ \vdots & \vdots & N_{i_S}^S(x_{i_F}^F) & \vdots \\ N_1^S(x_{n_F}^F) & N_2^S(x_{n_F}^F) & \cdots & N_{n_S}^S(x_{n_F}^F) \end{bmatrix}. \tag{3.29}$$

The shape function $N_{i_S}^S\left(x_{i_F}^F\right)$ belongs to the $i_S$-th node of a structure element and is evaluated at the $i_F$-th fluid node.

The resulting matrix $\mathbf{G}_{SF}$ interpolates the displacement $\boldsymbol{u}^S \in \mathbb{R}^{n_S}$ from the structure grid onto the fluid grid[5]

$$\boldsymbol{u}^F = \mathbf{G}_{SF}\, \boldsymbol{u}^S. \tag{3.30}$$

For a given fluid-structure grid combination, the interpolation matrix is constant. The interpolation matrix needs to be recomputed when the relative position of the points on each grid changes.

[5]This interpolation matrix is also used for the interpolation of the geometry from the structure grid onto the fluid grid.

### Pressure integration matrix $\mathbf{G}_{\text{int}}$

The integration of the fluid pressures, which are given at the fluid grid points, to forces, which are given at the structure grid points, is particularly critical. For one thing, the integration of the pressure must harmonize with the interpolation of the gap height, since the pressure reacts very sensitively to gap height changes. In addition, the pressure integration must result in a force vector corresponding to the finite element method.

In the finite element method, the content of the mass and the stiffness matrix, as well as the force vector, depend on the shape functions, see Subsection 2.2.2. Furthermore, the dimension of the force vector depends on the type of contact. The force vector has the dimension force per length for the line contact and the dimension force for the point contact. Based on Eq. (2.39) the force vector at the structure nodes $\boldsymbol{F} \in \mathbb{R}^{n_\mathrm{S}}$ can be computed as

$$\boldsymbol{F} = \begin{cases} \int_{x_0}^{x_{n_\mathrm{S}}} \boldsymbol{N}^\mathsf{T}(x)\, p(x)\, \mathrm{d}x, & \text{for a line contact, and} \\ \int_{x_0}^{x_{n_\mathrm{S}}} \boldsymbol{N}^\mathsf{T}(x)\, p(x)\, 2\pi\, x\, \mathrm{d}x, & \text{for an axisymmetric point contact,} \end{cases} \tag{3.31}$$

with the shape function vector $\boldsymbol{N}(x) \in \mathbb{R}^{1\times n_\mathrm{S}}$, and the pressure $p(x)$ at position $x$.[6] The structure domain $[x_0,\ x_{n_\mathrm{S}}]$ is split into small elements and the global coordinates $x$ are transformed into local coordinates $\xi$. The local coordinates in an element are $\xi \in [-1,\ 1]$. This allows for the solution of the integral using Gauss quadrature. To approximate the integral of a function using Gauss quadrature, the function is evaluated at unequally distributed quadrature points. A polynomial of order at most $(2\, n_\mathrm{Q} - 1)$ can be integrated exactly using $n_\mathrm{Q}$ quadrature points.

In local coordinates the quadratic shape functions, given via Eq. (3.28), simplify to

$$\begin{aligned} n_1(\xi) &= -\frac{\xi}{2}(1-\xi), \\ n_2(\xi) &= \frac{\xi}{2}(1+\xi), \text{ and} \\ n_3(\xi) &= 1-\xi^2, \end{aligned} \tag{3.32}$$

where the index 1, 2, and 3 indicate the left, right, and middle node, respectively. For a one-dimensional quadratic element, they can be expressed in matrix form as

$$\boldsymbol{N}_e(\xi) = \begin{bmatrix} n_1(\xi) & n_2(\xi) & n_3(\xi) \end{bmatrix}. \tag{3.33}$$

Figure 3.2 shows the quadratic shape functions of a one-dimensional element.

[6] For the axisymmetric point contact problem the spatial coordinate $x$ is the radial coordinate with origin at the center of the contact, i.e., $x = 0$.

The coordinate transformation from local to global coordinates is defined by

$$x = \chi_e(\xi) = \sum_{i_S=1}^{n_{en}} n_{i_S}(\xi)\, x^e_{i_S}, \tag{3.34}$$

where $x^e_{i_S}$ is the position of the $i_S$−th node of structure element $e$ and $n_{en}$ the number of nodes per element. The differential length d$x$ in Eq. (3.31) needs to be transformed into local coordinates. Here, one-dimensional quadratic elements with three nodes are used, i.e., $n_{en} = 3$. The corresponding relation becomes

$$\mathrm{d}x = \frac{\mathrm{d}\chi_e(\xi)}{\mathrm{d}\xi}\,\mathrm{d}\xi = \underbrace{\left(\frac{\mathrm{d}n_1(\xi)}{\mathrm{d}\xi}x^e_1 + \frac{\mathrm{d}n_2(\xi)}{\mathrm{d}\xi}x^e_2 + \frac{\mathrm{d}n_3(\xi)}{\mathrm{d}\xi}x^e_3\right)}_{J_e(\xi)}\mathrm{d}\xi. \tag{3.35}$$

For ease of notation, the derivative of the coordinate transformation will be abbreviated as $J_e(\xi)$ in the following.[7] Making use of relations (3.34) and (3.35) the integral in Eq. (3.31) can be written for each structure element $e$ as

$$\boldsymbol{F}_e = \begin{cases} \int_{-1}^{1} \boldsymbol{N}_e^{\mathsf{T}}(\xi)\, p(\chi_e(\xi))\, J_e(\xi)\,\mathrm{d}\xi, & \text{for a line contact, and} \\ \int_{-1}^{1} \boldsymbol{N}_e^{\mathsf{T}}(\xi)\, p(\chi_e(\xi))\, 2\pi\, \chi_e(\xi)\, J_e(\xi)\,\mathrm{d}\xi, & \text{for an axisymmetric point contact,} \end{cases} \tag{3.36}$$

where the local shape functions (3.32) are summarized in the vector $\boldsymbol{N}_e \in \mathbb{R}^{1\times 3}$ and $\boldsymbol{F}_e \in \mathbb{R}^{n^S_{en}}$. Applying Gauss quadrature to the integral in Eq. (3.36) yields

$$\boldsymbol{F}_e = \begin{cases} \sum_{i_Q=1}^{n_Q} w_{i_Q}\, \boldsymbol{N}_e^{\mathsf{T}}(\xi_{i_Q})\, p(\chi_e(\xi_{i_Q}))\, J_e(\xi_{i_Q}), & \text{for a line contact, and} \\ \sum_{i_Q=1}^{n_Q} w_{i_Q}\, \boldsymbol{N}_e^{\mathsf{T}}(\xi_{i_Q})\, p(\chi_e(\xi_{i_Q}))\, 2\pi\, \chi_e\left(\xi_{i_Q}\right)\, J_e(\xi_{i_Q}), & \text{for an axisym. point contact,} \end{cases} \tag{3.37}$$

with the weights $w_{i_Q}$, the quadrature points $\xi_{i_Q}$, and $i_Q = 1, 2, \ldots, n_Q$. The global force vector is constructed by applying this procedure to each structure element and subsequent assembly. A more detailed description of the construction of finite element force vectors can be found in [7, 67, 76].

The pressure integration is carried out using a precomputed matrix. The shape functions are evaluated at the quadrature points $\boldsymbol{N}_e^{\mathsf{T}}(\xi_{i_Q})$, the derivative of the coordinate transformation $J_e(\xi_{i_Q})$, and the weights $w_{i_Q}$ are known for a given structure grid. However, the pressure at the quadrature points depends on the pressure distribution on the fluid grid. Hence, the pressure needs

[7] The abbreviation "$J_e$" is chosen in analogy to the Jacobian matrix of the coordinate transformation, which is commonly used in the context of the finite element method.

to be interpolated from the fluid grid onto the quadrature points. To this end, an interpolation matrix, similar to the interpolation matrix $\mathbf{G}_{\mathrm{SF}}$, see Eq. (3.29), is constructed. However, different from $\mathbf{G}_{\mathrm{SF}}$, in this matrix, the shape functions are based on the fluid elements. The interpolation matrix is given by

$$\mathbf{G}^e_{\mathrm{FQ}} = \begin{bmatrix} N^{\mathrm{F}}_1(\chi_e(\xi_1)) & N^{\mathrm{F}}_2(\chi_e(\xi_1)) & \cdots & N^{\mathrm{F}}_{n_{\mathrm{F}}}(\chi_e(\xi_1)) \\ N^{\mathrm{F}}_1(\chi_e(\xi_2)) & N^{\mathrm{F}}_2(\chi_e(\xi_2)) & \cdots & N^{\mathrm{F}}_{n_{\mathrm{F}}}(\chi_e(\xi_2)) \\ \vdots & \vdots & N^{\mathrm{F}}_i(\chi_e(\xi_{i_{\mathrm{Q}}})) & \vdots \\ N^{\mathrm{F}}_1(\chi_e(\xi_{n_{\mathrm{Q}}})) & N^{\mathrm{F}}_2(\chi_e(\xi_{n_{\mathrm{Q}}})) & \cdots & N^{\mathrm{F}}_{n_{\mathrm{F}}}(\chi_e(\xi_{n_{\mathrm{Q}}})) \end{bmatrix}, \tag{3.38}$$

with dimensions $\mathbf{G}^e_{\mathrm{FQ}} \in \mathbb{R}^{n_{\mathrm{Q}} \times n_{\mathrm{F}}}$. The shape function $N^{\mathrm{F}}_{i_{\mathrm{F}}}(\chi_e(\xi_{i_{\mathrm{Q}}}))$ is based on a fluid element, it belongs to the $i_{\mathrm{F}}$–th fluid node and is evaluated at the $i_{\mathrm{Q}}$–th quadrature point in the $e$–th structure element.

The pre-evaluated quantities of the Gauss quadrature are summarized in quadrature matrix $\mathbf{G}^e_{\mathrm{QS}} \in \mathbb{R}^{n^{\mathrm{S}}_{\mathrm{en}} \times n_{\mathrm{Q}}}$, for a line contact as

$$\mathbf{G}^e_{\mathrm{QS}} = \begin{bmatrix} w_1 n_1(\xi_1)\, J_e(\xi_1) & w_2 n_1(\xi_2)\, J_e(\xi_2) & w_3 n_1(\xi_3)\, J_e(\xi_3) \\ w_1 n_2(\xi_1)\, J_e(\xi_1) & w_2 n_2(\xi_2)\, J_e(\xi_2) & w_3 n_2(\xi_3)\, J_e(\xi_3) \\ w_1 n_3(\xi_1)\, J_e(\xi_1) & w_2 n_3(\xi_2)\, J_e(\xi_2) & w_3 n_3(\xi_3)\, J_e(\xi_3) \end{bmatrix} \tag{3.39}$$

and for an axisymmetric point contact as

$$\mathbf{G}^e_{\mathrm{QS}} = 2\pi \begin{bmatrix} w_1 n_1(\xi_1)\, \chi_e(\xi_1)\; J_e(\xi_1) & w_2 n_1(\xi_2)\, \chi_e(\xi_2)\; J_e(\xi_2) & w_3 n_1(\xi_3)\, \chi_e(\xi_3)\; J_e(\xi_3) \\ w_1 n_2(\xi_1)\, \chi_e(\xi_1)\; J_e(\xi_1) & w_2 n_2(\xi_2)\, \chi_e(\xi_2)\; J_e(\xi_2) & w_3 n_2(\xi_3)\, \chi_e(\xi_3)\; J_e(\xi_3) \\ w_1 n_3(\xi_1)\, \chi_e(\xi_1)\; J_e(\xi_1) & w_2 n_3(\xi_2)\, \chi_e(\xi_2)\; J_e(\xi_2) & w_3 n_3(\xi_3)\, \chi_e(\xi_3)\; J_e(\xi_3) \end{bmatrix}. \tag{3.40}$$

The locations of the quadrature points are given as $\xi_1 = -\sqrt{(3/5)}$, $\xi_2 = 0$, and $\xi_3 = \sqrt{(3/5)}$. The corresponding weights are $w_1 = 5/9$, $w_2 = 8/9$, and $w_3 = 5/9$, respectively. Weights and locations for various numbers of quadrature points can be found in finite element textbooks such as [7].

Combining the pressure interpolation matrix $\mathbf{G}^e_{\mathrm{FQ}}$ with the quadrature matrix $\mathbf{G}^e_{\mathrm{QS}}$ gives the element pressure integration matrix

$$\mathbf{G}^e_{\mathrm{int}} = \mathbf{G}^e_{\mathrm{QS}}\, \mathbf{G}^e_{\mathrm{FQ}}, \tag{3.41}$$

with which the element force vector can be computed as

$$\boldsymbol{F}_e = \mathbf{G}^e_{\mathrm{int}}\, \boldsymbol{p}, \tag{3.42}$$

where $\boldsymbol{p} \in \mathbb{R}^{n_\mathrm{F}}$ contains the pressure values at all fluid nodes. The pressure integration matrix $\mathbf{G}_\mathrm{int} \in \mathbb{R}^{n_\mathrm{S} \times n_\mathrm{F}}$ is constructed by looping over all structure elements. The pressure integration matrix is constant for a given fluid-structure grid combination. It needs to be recomputed when the relative position of the grids changes.

### 3.5.3 Interpolation onto the computational grid

The coupling matrices map the field variables of the coupled problem onto the computational grid $\boldsymbol{x}^\mathrm{F}$, see Fig. 3.1. The pressure integration matrix $\mathbf{G}_\mathrm{int} \in \mathbb{R}^{n_\mathrm{S} \times n_\mathrm{F}}$, see Eq. (3.41), integrates the pressures $\boldsymbol{p} \in \mathbb{R}^{n_\mathrm{F}}$ from the fluid grid to forces $\boldsymbol{F} \in \mathbb{R}^{n_\mathrm{S}}$ on the structure grid, using the relation

$$\boldsymbol{F} = \mathbf{G}_\mathrm{int}\, \boldsymbol{p}. \tag{3.43}$$

With Eqs. (3.23) and (3.25) the displacements $\boldsymbol{u} \in \mathbb{R}^{n_\mathrm{S}}$ are computed on the structure grid. Using the interpolation matrix $\mathbf{G}_\mathrm{SF} \in \mathbb{R}^{n_\mathrm{F} \times n_\mathrm{S}}$, see Eq. (3.29), the displacements are interpolated onto the fluid grid.

The gap displacements $\boldsymbol{\delta} \in \mathbb{R}^{n_\mathrm{F}}$ are computed as the sum of both bodies' displacements, i.e.,

$$\boldsymbol{\delta} = \mathbf{G}_{\mathrm{SF}\,[1]}\, \boldsymbol{u}_{[1]} + \mathbf{G}_{\mathrm{SF}\,[2]}\, \boldsymbol{u}_{[2]}, \tag{3.44}$$

where the subscripts [1] and [2] indicate the lower and the upper body, respectively.

The deformed gap height is computed as

$$\boldsymbol{h} = \boldsymbol{h}_\mathrm{geo} + \boldsymbol{\delta}, \tag{3.45}$$

where the undeformed gap height is given via

$$\boldsymbol{h}_\mathrm{geo} = \mathbf{G}_{\mathrm{SF}\,[1]}\, \boldsymbol{z}_{\mathrm{geo},[1]} + \mathbf{G}_{\mathrm{SF}\,[2]}\, \boldsymbol{z}_{\mathrm{geo},[2]}, \tag{3.46}$$

with the coordinates of the undeformed geometry $\boldsymbol{z}_\mathrm{geo} \in \mathbb{R}^{n_\mathrm{S}}$. The deformed gap height is nondimensionalized using relations (2.19) to give

$$\boldsymbol{H} = \boldsymbol{H}_\mathrm{geo} + \bar{\boldsymbol{\delta}}. \tag{3.47}$$

Now all field variables are given on the computational grid $\boldsymbol{x}^\mathrm{F}$.

### 3.5.4 Summary

The presented fluid-structure coupling concludes the first part of this chapter, which outlined the construction of the structural model and its coupling. Table 3.1 summarizes the steps taken.

**Table 3.1:** This table summarizes the preparation process of the structural model, as outlined in the first sections of this chapter. It shows the output of each step. Thereby, it illustrates the reduction steps of the structural model and the preparation of the coupling.

**Finite element model, see Subsection 2.2.2**

The bodies are created in a CAE software and discretized using finite elements. The system is of size $n_u$.

$$\mathbf{M}\ddot{\boldsymbol{u}} + \mathbf{K}\boldsymbol{u} = \boldsymbol{F} \tag{3.1}$$

**Reduced order model, see Section 2.3**

From the finite element model, a reduced order model is created using component mode synthesis. The system is of size $\widetilde{n}_u$ whereby usually $\widetilde{n}_u \ll n_u$ holds.

$$\tilde{\mathbf{M}}\ddot{\tilde{\boldsymbol{u}}} + \tilde{\mathbf{K}}\tilde{\boldsymbol{u}} = \tilde{\boldsymbol{F}} \tag{3.2}$$

**Transformation into the modal space, see Section 3.2**

Transformation into the modal space speeds up the computation and allows for damping individual modes. The system size remains unchanged, i.e., $n_q = \widetilde{n}_u$.

$$\begin{aligned} \ddot{\boldsymbol{q}} + \overline{\mathbf{K}}\,\boldsymbol{q} &= \bar{\boldsymbol{\Phi}}^{\mathsf{T}}\,\tilde{\boldsymbol{F}} \\ \tilde{\boldsymbol{u}} &= \bar{\boldsymbol{\Phi}}\,\boldsymbol{q} \end{aligned} \tag{3.5}$$

**Modified state-space representation, see Sections 3.3 and 3.4**

The modified state-space representation reduces the system size and explicitly expresses the displacements as a function of the force. The matrices depend on the treatment approach.

$$\begin{aligned} \boldsymbol{Y}_1^{jk} &= \left(\mathbf{A}_{21} + \frac{1}{\Delta t_j\, a_{ii}}\left(\mathbf{A}_{22} - \frac{1}{\Delta t_j\, a_{ii}}\mathbf{I}\right)\right)^{-1} \\ &\quad \left(\frac{1}{\Delta t_j\, a_{ii}}\left(\mathbf{A}_{22} - \frac{1}{\Delta t_j\, a_{ii}}\mathbf{I}\right) \boldsymbol{S}_1^{jk} - \frac{1}{\Delta t_j\, a_{ii}}\boldsymbol{S}_2^{jk} - \mathbf{B}_2\,\tilde{\boldsymbol{F}}^{jk}\right) \end{aligned} \tag{3.23}$$

$$\tilde{\boldsymbol{u}}^{jk} = \mathbf{C}\,\boldsymbol{Y}_1^{jk} + \mathbf{D}\,\tilde{\boldsymbol{F}}^{jk} \tag{3.25}$$

**Transformation to the computational grid, see Section 3.5**

- Equations (3.23) and (3.25) give the body's displacement as a function of the force. The force and the displacement are given on the structure grid $\boldsymbol{x}^{\mathrm{S}} \in \mathbb{R}^{n_{\mathrm{S}}}$.
- The interpolation matrix $\mathbf{G}_{\mathrm{SF}} \in \mathbb{R}^{n_{\mathrm{F}} \times n_{\mathrm{S}}}$ interpolates the structural displacements to displacements on the computational grid via
$$\boldsymbol{u}^{\mathrm{F}} = \mathbf{G}_{\mathrm{SF}}\,\boldsymbol{u}^{\mathrm{S}}. \tag{3.30}$$
- The pressure integration matrix $\mathbf{G}_{\mathrm{int}} \in \mathbb{R}^{n_{\mathrm{S}} \times n_{\mathrm{F}}}$ integrates the fluid pressures to the structural forces via
$$\boldsymbol{F} = \mathbf{G}_{\mathrm{int}}\,\boldsymbol{p}. \tag{3.43}$$

## 3.6 Solution of the nonlinear system of equations

Based on the extended structural model and the presented fluid-structure coupling, this section shows the solution process of the resulting coupled problem. It outlines the build-up of the system of equations and the iterative solution procedure.

### Setting up the system of equations

The problem is coupled using a monolithic coupling approach. The nonlinear system of equations describing the coupled problem is given as

$$\boldsymbol{f}\left(\boldsymbol{P}^{jk},\bar{\boldsymbol{\delta}}^{jk}\right)=\left\{\begin{array}{c}\boldsymbol{f}_p\left(\boldsymbol{P}^{jk},\bar{\boldsymbol{\delta}}^{jk}\right)\\ \boldsymbol{f}_\delta\left(\boldsymbol{P}^{jk},\bar{\boldsymbol{\delta}}^{jk}\right)\end{array}\right\}=\mathbf{0}, \tag{3.48}$$

with the differentiable function $\boldsymbol{f}\in\mathbb{R}^{n_z}$ and the size of the system $n_z = n_p + n_\mathrm{F}$, where $n_p$ is the number of unknown pressures and $n_\mathrm{F}$ the dimension of the computational grid. The problem is made up of two parts: the nonlinear fluid model and the linear structural model.

The fluid model is introduced into the coupled problem through the discretized Reynolds Equation (2.21). The Reynolds Equation is evaluated at each grid point where the pressure is unknown. The results are collected in the vector $\boldsymbol{f}_p\left(\boldsymbol{P}^{jk},\bar{\boldsymbol{\delta}}^{jk}\right)\in\mathbb{R}^{n_p}$. The vectors $\boldsymbol{P}^{jk}\in\mathbb{R}^{n_\mathrm{F}}$ and $\bar{\boldsymbol{\delta}}^{jk}\in\mathbb{R}^{n_\mathrm{F}}$ contain the dimensionless pressures and gap displacements, respectively.[8]

The structural model is implicitly introduced into the system of equations through the gap displacements. The gap displacements can be expressed in nondimensionalized and discrete form as

$$\boldsymbol{f}_\delta=\bar{\boldsymbol{\delta}}-\left(\mathbf{G}_{\mathrm{SF}\,[1]}\,\boldsymbol{U}_{[1]}+\mathbf{G}_{\mathrm{SF}\,[2]}\,\boldsymbol{U}_{[2]}\right)=\mathbf{0}, \tag{3.49}$$

with the differentiable function $\boldsymbol{f}_\delta\in\mathbb{R}^{n_\mathrm{F}}$ , the interpolation matrix $\mathbf{G}_\mathrm{SF}\in\mathbb{R}^{n_\mathrm{F}\times n_\mathrm{S}}$, see (3.29), and the dimensionless structural displacements $\boldsymbol{U}\in\mathbb{R}^{n_\mathrm{s}}$. The latter two are computed for the lower body and for the upper body, denoted in Eq. (3.49) by index [1] and [2], respectively.

Here, the linearity of the structural model utilized. The linearity allows expressing the dimensionless structural displacements $\boldsymbol{U}\in\mathbb{R}^{n_\mathrm{F}}$ as a function of the dimensionless pressures $\boldsymbol{P}^{jk}\in\mathbb{R}^{n_\mathrm{F}}$. The displacements are computed via Eqs. (3.23) and (3.25) on the structure grid and interpolated after that to give the gap displacement on the computational grid. This approach yields two significant advantages:

[8]The superscript $jk$ indicates the current stage of the time integration. For ease of notation, it is omitted in the following.

- The displacements of both bodies are introduced into the system of equations via the gap displacements $\bar{\boldsymbol{\delta}}$ rather than via the displacement vector of each body. This approach reduces the system's dimension, which results in reduced computational costs for the solution of the system of equations.
- The computations related to the structural model are carried out on the lower dimensional structural grid.

### Solving the coupled problem

A solution of the nonlinear system of equations (3.48) is approximated iteratively using the Newton-Raphson method. To this end, the system of equations (3.48) is linearized at $z^{(m)} \in \mathbb{R}^{n_z}$, where the vector

$$z^{(m)} = \left\{ \begin{array}{c} \boldsymbol{P}^{(m)} \\ \bar{\boldsymbol{\delta}}^{(m)} \end{array} \right\} \tag{3.50}$$

contains the unknowns and the superscript $m$ denotes the $m-$th iteration of the Newton-Raphson method. The linearized function is set to be zero and the increment $\Delta z^{(m)} = z - z^{(m)}$ is introduced

$$\mathbf{0} \stackrel{!}{=} \boldsymbol{f}\left(z^{(m)}\right) - \mathbf{J}_f\left(z^{(m)}\right) \Delta z^{(m)}, \tag{3.51}$$

where $\mathbf{J}_f\left(z^{(m)}\right) = \left.\frac{\partial f}{\partial z}\right|_{z^{(m)}}$ is the Jacobian matrix of the nonlinear function. With the increment

$$\Delta z^{(m)} = -\left(\mathbf{J}_f\left(z^{(m)}\right)\right)^{-1} \boldsymbol{f}\left(z^{(m)}\right), \tag{3.52}$$

and the under-relaxation factor $\omega^{(m)}$ a new approximation of the solution is computed

$$z^{(m+1)} = z^{(m)} + \omega^{(m)} \Delta z^{(m)}. \tag{3.53}$$

This process is repeated until a predefined convergence criterion is fulfilled, such as $||\boldsymbol{f}(z^{(m)})|| < tol$, where $tol$ is an arbitrarily small number.

The under-relaxation factor $\omega^{(m)}$ is introduced to stabilize the numerical method. However, a too small choice of $\omega^{(m)}$ slows down convergence. Hence, different methods exist to choose $\omega^{(m)}$ depending on the phase of the solution adaptively. In this work, Aitken-method is used to determine an under-relaxation factor, see [97, 112].

To speed up the solution of the linear system (3.51), it is worthwhile to study the matrix population of the Jacobian matrix. In a straightforward approach an inversion of the Jacobian matrix, which is of size $n_z$, would be necessary to solve the linear system. However, the Jacobian is a block matrix consisting of four blocks

$$\mathbf{J}_f = \begin{bmatrix} \mathbf{J}_{pp} & \mathbf{J}_{pd} \\ \mathbf{J}_{dp} & \mathbf{J}_{dd} \end{bmatrix}, \tag{3.54}$$

where $\mathbf{J}_{pp} = \partial \boldsymbol{f}_p / \partial \boldsymbol{P}$ and $\mathbf{J}_{pd} = \partial \boldsymbol{f}_p / \partial \bar{\boldsymbol{\delta}}$ are sparse matrices, $\mathbf{J}_{dp} = \partial \boldsymbol{f}_\delta / \partial \boldsymbol{P}$ is fully populated and $\mathbf{J}_{dd} = \partial \boldsymbol{f}_\delta / \partial \bar{\boldsymbol{\delta}} = \mathbf{I}$ is the identity matrix.

The structure of the Jacobian matrix indicates that the application of the Schur complement to solve the linear system of equations is particularly helpful.[9] Writing the system of equations (3.51) block-wise yields

$$\mathbf{J}_{pp}\, \Delta \boldsymbol{P} + \mathbf{J}_{pd}\, \Delta \bar{\boldsymbol{\delta}} = \boldsymbol{f}_p, \text{ and} \tag{3.55a}$$

$$\mathbf{J}_{dp}\, \Delta \boldsymbol{P} + \mathbf{J}_{dd}\, \Delta \bar{\boldsymbol{\delta}} = \boldsymbol{f}_\delta. \tag{3.55b}$$

Multiplication of Eq. (3.55b) with $\mathbf{J}_{pd}\,\mathbf{J}_{dd}^{-1}$ and substraction from Eq. (3.55a) gives

$$\left(\mathbf{J}_{pp} - \mathbf{J}_{pd}\,\mathbf{J}_{dd}^{-1}\,\mathbf{J}_{dp}\right)\, \Delta \boldsymbol{P} = \boldsymbol{f}_p - \mathbf{J}_{pd}\,\mathbf{J}_{dd}^{-1}\,\boldsymbol{f}_\delta, \tag{3.56}$$

which can be solved for $\Delta \boldsymbol{P}$, which can after that be inserted into Eq. (3.55b) to solve for $\Delta \bar{\boldsymbol{\delta}}$. This approach is particularly useful since $\mathbf{J}_{dd}$ is the identity matrix and no inversion is necessary, as $\mathbf{J}_{dd} = \mathbf{J}_{dd}^{-1} = \mathbf{I}$. Using this approach, only an inversion of a matrix of size $n_p$ is necessary to solve the linear system, which can save significant computational time because $n_p \approx n_z/2$ .

The convergence behavior of the Newton-Raphson method depends on the accuracy of the Jacobian matrix and on the proximity of the starting value to the solution. Therefore, the Jacobian matrix is calculated based on analytical derivatives and recomputed in every iteration. Furthermore, the starting value of each stage is based on a linear extrapolation of the previous stage.

## 3.7 Model order reduction of the nonlinear system

Sections 3.1 through 3.6 provide a framework to solve the EHD contact problem with an extended structural model. This section outlines the application of model order reduction (MOR) to the complete extended EHD contact model. This procedure is used to construct a reduced order model (ROM), which can significantly reduce the computational effort.

[9] Further applications of the Schur complement can be found, for example, in [139].

While the solution steps in Sections 3.1 through 3.6 are needed to compute a solution for the coupled problem, this method aims to further reduce the computational effort. However, the application of MOR is not useful in all applications. Whether the use of MOR makes sense depends to a large extent on the application. If the similarity of the solutions within the solution space is high and many solutions within the solution space have to be computed, the application of MOR usually makes sense.

The procedure described below is based on [93]. It uses Proper Orthogonal Decomposition (POD) to construct a projection and then further reduces the size of the system using a system approximation.

### Reducing the system size by projection

Subsection 2.3.1 outlines the concept of MOR by projection. The aim is to approximate the behavior of the full system's behavior using a lower dimensional space. The solution given on the lower dimensional space is the solution of the reduced order model (ROM). The ROM shall approximate the essential input-output behavior of the original model with a much smaller number of degrees of freedom, i.e., $\widetilde{n}_z \ll n_z$, where $\widetilde{n}_z$ is the number of unknowns of the ROM and $n_z$ is the number of unknowns of the full model. To project the full system onto the reduced space, a projection basis $\mathbf{V}_z \in \mathbb{R}^{n_z \times \widetilde{n}_z}$ needs to be constructed. The projection basis of the full system $\mathbf{V}_z$ is constructed from the individual projection bases of the different field variables. Here, projection bases are constructed for the fluid part, i.e., the pressure $\boldsymbol{P} \in \mathbb{R}^{n_p}$, and for the structural part, i.e., the gap displacement $\bar{\boldsymbol{\delta}} \in \mathbb{R}^{n_\delta}$, where $n_p$ and $n_\delta$ are the number of unknown pressures and gap displacements, respectively, with $n_\delta = n_F$. Subsequently, the projection basis of the full system is constructed as

$$\mathbf{V}_z = \operatorname{diag}\left(\mathbf{V}_p, \mathbf{V}_\delta\right) \in \mathbb{R}^{n_z \times \widetilde{n}_z}, \tag{3.57}$$

with $n_z = n_p + n_\delta$ and $\widetilde{n}_z = \widetilde{n}_p + \widetilde{n}_\delta$ analogously.

In the following, the procedure to construct a projection basis is described exemplarily for the field variable *pressure*.

*Constructing the projection basis*

The projection basis is derived using so-called trainings. The trainings are typically solutions of the full system (3.48) for certain parameters and boundary conditions. The training solutions are collected column-wise in the snapshot matrix $\mathbf{S} \in \mathbb{R}^{n_p \times n_s}$, with the number of snapshots $n_s$. In transient problems all time steps of the transient solutions for the chosen parameter combinations are saved in the snapshot matrix.[10] The choice of trainings has a great impact on the range of parameters and boundary conditions, which the ROM will be able to approximate.

The Proper Orthogonal Decomposition (POD) method is used to create the projection basis based on the snapshot matrix.[11] To this end, a singular value decomposition (SVD) of the snapshot matrix is carried out. The SVD of $\mathbf{S}$ yields

$$\mathbf{S} = \mathbf{U}\,\boldsymbol{\Sigma}\,\mathbf{W}^{\mathsf{T}}, \tag{3.58}$$

with the left singular matrix $\mathbf{U}$ and the right singular matrix $\mathbf{W}$. The diagonal matrix $\boldsymbol{\Sigma}$ contains the singular values $\sigma_i$ sorted in decreasing order of magnitude, with $i = 1,\ 2,\ \ldots,\ \min\left(n_p,\ n_s\right)$. All three matrices are orthogonal matrices.

The basis for the solution space of the ROM is given by the vectors of the left singular matrix $\mathbf{U}$

$$\mathbf{V}_p \subseteq \mathbf{U} \tag{3.59}$$

with $\mathbf{V}_p \in \mathbb{R}^{n_p \times \widetilde{n}_p}$. The number of considered vectors $\widetilde{n}_p$ is defined with help of the singular values $\sigma_i$. They are used as an indicator of the information content of a given singular vector, which is given as the $i$-th column of the singular matrix $\mathbf{U}(:, i)$. To this end, information loss values are computed as

$$l(k) = 1 - \frac{\sum_{i=1}^{k} \sigma_i}{\sum_{i=1}^{\min(n_p, n_s)} \sigma_i}. \tag{3.60}$$

The loss values indicate how much relative information is lost by neglecting all singular vectors $\mathbf{U}(:, i)$ for which $i > k$ holds. Since the singular values are sorted in descending order, the loss values $l(k)$ are monotonically decreasing. Therefore, for a given loss value $l_{\max}$ the first

$$\widetilde{n}_p := \underset{j \in \mathbb{N}}{\operatorname{argmin}} \left(l(j) \leq l_{\max}\right) \tag{3.61}$$

singular vectors are used to construct the projection basis

$$\mathbf{V}_p = \mathbf{U}(:,\ 1 : \widetilde{n}_p) \in \mathbb{R}^{n_p \times \widetilde{n}_p}. \tag{3.62}$$

The solution of the full model can be approximated as $\boldsymbol{P} \approx \mathbf{V}_p \tilde{\boldsymbol{P}}$ with vector $\tilde{\boldsymbol{P}} \in \mathbb{R}^{\widetilde{n}_p}$.

[10] In the case of SDIRK methods, all stage values are saved in the snapshot matrix, see Eq. (2.74).

[11] For further detail on POD see, for example, [15] and [16].

Above, the procedure is outlined exemplarily for the pressure $\boldsymbol{P} \in \mathbb{R}^{n_P}$. The same procedure is carried out for the gap displacement $\bar{\boldsymbol{\delta}} \in \mathbb{R}^{n_\delta}$. Subsequently, a projection basis of the full system is constructed according to Eq. (3.57).

*Projecting the full system onto the reduced space*

According to Eqs. (2.45) through (2.49) the full system (3.48) is projected onto the lower dimensional space with the global projection matrix $\mathbf{V}_z$. The nonlinear function (3.48) and its Jacobian matrix become

$$\tilde{\boldsymbol{f}}\,(\mathbf{V}_z\,\tilde{z}) = \left\{ \begin{array}{c} \mathbf{V}^{\mathsf{T}}{}_P\,\boldsymbol{f}_P\,(\mathbf{V}_z\,\tilde{z}) \\ \mathbf{V}^{\mathsf{T}}{}_\delta\,\boldsymbol{f}_\delta\,(\mathbf{V}_z\,\tilde{z}) \end{array} \right\}, \text{ and} \tag{3.63}$$

$$\tilde{\mathbf{J}}_f\,(\mathbf{V}_z\,\tilde{z}) = \left[ \begin{array}{c} \mathbf{V}^{\mathsf{T}}{}_P\,\mathbf{J}_{f_P}\,\mathbf{V}_z \\ \mathbf{V}^{\mathsf{T}}{}_\delta\,\mathbf{J}_{f_\delta}\,\mathbf{V}_z \end{array} \right], \tag{3.64}$$

with $\tilde{\boldsymbol{f}} \in \mathbb{R}^{\tilde{n}_z}$ and $\tilde{\mathbf{J}}_f \in \mathbb{R}^{\tilde{n}_z \times \tilde{n}_z}$.

Consequently, the formula for the Newton-Raphson method, see Eq. (3.51), becomes

$$\mathbf{0} \overset{!}{=} \mathbf{V}^{\mathsf{T}}{}_z\,\boldsymbol{f}\,(\,\mathbf{V}_z\,\tilde{z}\,) + \mathbf{V}^{\mathsf{T}}{}_z\,\mathbf{J}_f(\mathbf{V}_z\,\tilde{z})\,\mathbf{V}_z\,\Delta\,\tilde{z}, \tag{3.65}$$

where the increment in the reduced space is computed as

$$\Delta\tilde{z} = -\left(\tilde{\mathbf{J}}_f\,(\mathbf{V}_z\tilde{z})\right)^{-1}\,\tilde{\boldsymbol{f}}\,(\mathbf{V}_z\tilde{z})\,. \tag{3.66}$$

This projection reduces the size of the matrix that needs to be inverted in each iteration of the Newton-Raphson method from $n_z$ to $\tilde{n}_z$. Depending on the difference between $\tilde{n}_z$ and $n_z$, this reduction can significantly decrease the computational effort of the solution procedure.

### Reducing the number of state variables

The projection onto the lower dimensional space reduces the size of the nonlinear system from $n_z$ to $\tilde{n}_z$. However, the nonlinear function $\boldsymbol{f} \in \mathbb{R}^{n_z}$ and its Jacobian matrix $\mathbf{J}_f \in \mathbb{R}^{n_z \times n_z}$ still need to be computed on the full system level, which is of size $n_z$, see Eqs. (3.63) and (3.64). The computation of the functions on full system level takes a significant amount of time. To bypass this, Chaturantabut and Sorensen [17, 18] proposed a method to evaluate only certain points of the nonlinear function and interpolate between these points. A drastic reduction of the number of evaluation points without a significant decrease in solution quality was shown. Carlberg [13, 14] extended the method by introducing different interpolations for both the system function and its Jacobian matrix. Maier [95] extended and applied this method to the EHD contact problem.

The goal is to reduce the cost of the evalutation of the system functions. The majority of the cost stems from the construction of the fluid part, i.e., Reynolds Equation $\boldsymbol{f}_p \in \mathbb{R}^{n_p}$ and its Jacobian matrix $\mathbf{J}_{f_p} \in \mathbb{R}^{n_p \times n_z}$. The idea is to identify $\widetilde{N}_p$ evaluation points which are sufficient to characterize the state of the nonlinear system, such that the system functions only need to be computed at the evaluation points. The indices of the evaluation points are summarized in the set $\mathcal{I}$.

Similar to the construction of the projection basis, snapshot matrices are created. The snapshot matrices $\mathbf{S}_f$ and $\mathbf{S}_J$ contain the nonlinear function $\boldsymbol{f}_p^{(m)}$ and its weighted Jacobian $\mathbf{J}_{f_p}^{(m)} z^{(m)}$ of each Newton-Raphson iteration of all time steps and all parameter combinations, respectively.[12] The SVD of the snapshot matrices yield

$$\mathbf{S}_f = \mathbf{U}_f \, \boldsymbol{\Sigma}_f \, \mathbf{W}^{\mathsf{T}}{}_f, \text{ and} \tag{3.67}$$

$$\mathbf{S}_J = \mathbf{U}_J \, \boldsymbol{\Sigma}_J \, \mathbf{W}^{\mathsf{T}}{}_J, \tag{3.68}$$

with the left singular matrix $\mathbf{U}$, the right singular matrix $\mathbf{W}$ and the diagonal matrix $\boldsymbol{\Sigma}$, analogously to Eq. (3.58). Based on the prescribed number of evaluation points $\widetilde{N}_p$, the matrices $\boldsymbol{\Phi}_f \in \mathbb{R}^{n_p \times \widetilde{N}_p}$, and $\boldsymbol{\Phi}_J \in \mathbb{R}^{n_p \times \widetilde{N}_p}$ are created from the first $\widetilde{N}_p$ columns of the left singular matrices $\mathbf{U}_f$ and $\mathbf{U}_J$, respectively,

$$\boldsymbol{\Phi}_f = \mathbf{U}_f \left(:, \; 1 : \widetilde{N}_p\right) \in \mathbb{R}^{n_p \times \widetilde{N}_p}, \text{ and} \tag{3.69}$$

$$\boldsymbol{\Phi}_J = \mathbf{U}_J \left(:, \; 1 : \widetilde{N}_p\right) \in \mathbb{R}^{n_p \times \widetilde{N}_p}. \tag{3.70}$$

Using matrices $\boldsymbol{\Phi}_f$ and $\boldsymbol{\Phi}_J$ and the number of evaluation points $\widetilde{N}_p$, the set $\mathcal{I}$ can be computed. Figure 3.3 outlines the algorithm to do so.

To compute $\boldsymbol{f}_p$ and $\mathbf{J}_{f_p}$ at the evaluation points given by the set $\mathcal{I}$, also some neighboring points need to be known.

Due to the discretization, the values of the Reynolds Equation at the points of set $\mathcal{I}$ are dependent on the values at their neighboring grid points. For this reason the additional set $\mathcal{J}$ is introduced. Set $\mathcal{J}$ contains all points needed to compute $\boldsymbol{f}_p$ and $\mathbf{J}_{f_p}$ at the evaluation points. Set $\mathcal{J}$ is defined as

$$\mathcal{J} := \left\{ j \in \{1, \ldots, n_z\} : \exists l \in \mathcal{I} \text{ with } \mathbf{J}_{f_p}(l, j) \neq 0 \right\}, \tag{3.71}$$

where $\hat{N}_p = |\mathcal{J}|$ gives the number of elements of the set $\mathcal{J}$, i.e., the number of evaluation points needed.

[12] Note that this procedure is only carried out for the Reynolds Equation. The structural equations are linear and can thus be obtained with little numerical cost.

Input: $\mathbf{\Phi}_f$, $\mathbf{\Phi}_J$, $\widetilde{N}_p$
Output: $\mathcal{I}$

$$
\begin{aligned}
i &= \underset{j=1,\,2,\,\ldots,\,n_p}{\operatorname{argmax}} \left[ \left(\mathbf{\Phi}_f(j,1)\right)^2 + \left(\mathbf{\Phi}_J(j,1)\right)^2 \right] \\
\mathcal{I} &= \{i\}
\end{aligned}
$$

for $k = 2 : \widetilde{N}_p$

$$
\begin{aligned}
\boldsymbol{a}_f &= \underset{\boldsymbol{a}_f \in \mathbb{R}^{k-1}}{\operatorname{argmin}} \left\| \mathbf{\Phi}_f(\mathcal{I},\ 1:k-1)\,\boldsymbol{a}_f - \mathbf{\Phi}_f(\mathcal{I},k) \right\|_2 \\
\boldsymbol{a}_J &= \underset{\boldsymbol{a}_J \in \mathbb{R}^{k-1}}{\operatorname{argmin}} \left\| \mathbf{\Phi}_J(\mathcal{I},\ 1:k-1)\,\boldsymbol{a}_J - \mathbf{\Phi}_J(\mathcal{I},k) \right\|_2 \\
\boldsymbol{\varphi}_f &= \mathbf{\Phi}_f(:,k) - \mathbf{\Phi}_f(:,\ 1:k-1)\,\boldsymbol{a}_f \\
\boldsymbol{\varphi}_J &= \mathbf{\Phi}_J(:,k) - \mathbf{\Phi}_J(:,\ 1:k-1)\,\boldsymbol{a}_J \\
i &= \underset{j=1,\,2,\,\ldots,\,n_p \setminus \mathcal{I}}{\operatorname{argmax}} \left[ \left(\boldsymbol{\varphi}_f(j)\right)^2 + \left(\boldsymbol{\varphi}_J(j)\right)^2 \right] \\
\mathcal{I} &= \mathcal{I} \cup \{i\}
\end{aligned}
$$

end

**Figure 3.3:** Algorithm to determine $\mathcal{I}$, adapted from [95]. The set $\mathcal{I}$ contains the evaluation points.

The nonlinear function and its Jacobian matrix are now constructed only at the evaluation points, hence, $\boldsymbol{f}_p(\mathcal{I}) \in \mathbb{R}^{\widetilde{N}_p}$ and $\mathbf{J}_{f_p}(\mathcal{I},\mathcal{J}) \in \mathbb{R}^{\widetilde{N}_p \times \hat{N}_p}$. The system functions of the ROM can be expressed as

$$
\hat{\boldsymbol{f}}(\mathbf{V}_z\,\tilde{z}) = \left\{ \begin{array}{c} \left(\mathbf{V}_p(\mathcal{I},:)\right)^{\mathsf{T}} \boldsymbol{f}_p\left(\mathbf{V}_z(\mathcal{I},:)\,\tilde{z}\right) \\ \mathbf{V}^{\mathsf{T}}{}_{\delta}\,\boldsymbol{f}_{\delta}\left(\mathbf{V}_z\,\tilde{z}\right) \end{array} \right\}, \text{ and} \tag{3.72}
$$

$$
\hat{\mathbf{J}}_f(\mathbf{V}_z\,\tilde{z}) = \left[ \begin{array}{c} \left(\mathbf{V}_p(\mathcal{I},:)\right)^{\mathsf{T}} \mathbf{J}_{f_p}(\mathcal{I},\mathcal{J})\,\mathbf{V}_z(\mathcal{J},:) \\ \mathbf{V}^{\mathsf{T}}{}_{\delta}\,\mathbf{J}_{f_\delta}\,\mathbf{V}_z \end{array} \right]. \tag{3.73}
$$

The system approximation only reduces the points at which the nonlinear function $\boldsymbol{f}$ and its Jacobian matrix $\mathbf{J}_f$ need to be evaluated on full system level. The projection matrix $\mathbf{V}_z$ remains unchanged. Therefore, the size of the reduced system also remains unchanged, i.e., $\hat{\boldsymbol{f}} \in \mathbb{R}^{\widetilde{n}_z}$ and $\hat{\mathbf{J}}_f \in \mathbb{R}^{\widetilde{n}_z \times \widetilde{n}_z}$. However, after the system approximation, during computation with the ROM, no operations need to be performed on the full system level. The reduced number of full system evaluation speeds up the overall procedure.

# 4 Results

This chapter investigates the developed solution procedure for the extended EHD contact model. Different example problems are studied, in order to examine individual aspects of the procedure.

The shock loading of an EHD line contact is used to investigate the proposed time integration procedure and to draw a comparison with conventional procedures.

The dry impact problem of a plane valve-like geometry is used to investigate the extended structural model. Aspects of the reduction procedure and the occurrence of spurious oscillations in reduced order models are discussed.

Various problems based on the axisymmetric lubricated impact problem are studied to examine the complete extended EHD contact model. The first problem studied is the impact of a sphere on a lubricated plate. Based on this problem, the model's overall behavior is validated by comparison with literature results. The second problem investigates an application where structural inertia effects influence the contact behavior, i.e., the lubricated impact of a valve-like geometry. The last problem of Section 4.3 explores an efficient solution approach for multiple successive impacts. Here, the model order reduction of the complete nonlinear coupled problem is shown.

## 4.1 Shock loading of a lubricated contact

EHD contact solvers commonly use the Backward Euler method or the second-order finite difference method for time integration, see, e.g., [47, 48] or [93, 128, 134–136], respectively. Different from the conventional approaches, this work uses SDIRK methods for time integration. These methods provide a stable and efficient solution procedure for stiff problems. However, they have hardly been explored in the EHD context. Therefore, this section examines their suitability for EHD contact problems. The following study investigates the temporal characteristics of the fluid part of the coupled problem in isolation. Hence, for this problem, the conventional EHD line contact model, which is widely used for EHD contact problems, is selected. In this model, structural inertia is neglected and the solid's deformation is modeled via the half-space approach. Hence, only the Reynolds Equation has time-dependent terms.

The problem of shock loading and unloading for a highly loaded contact is used as an example problem. This problem was used by Goodyer [47, 48] to study the temporal characteristics of the EHD contact and is also used here, however, to study a different time integration approach. The problem is particularly suitable since it consists of highly dynamic phases of rapid load change and also of phases of steady-state operation.

The first subsection states the example problem. Subsection 4.1.2 compares the proposed time integration method with conventionally used methods. Subsection 4.1.3 studies the adaptive time stepping approach and, furthermore, extends it in order to consider rapid load changes during time step size selection.

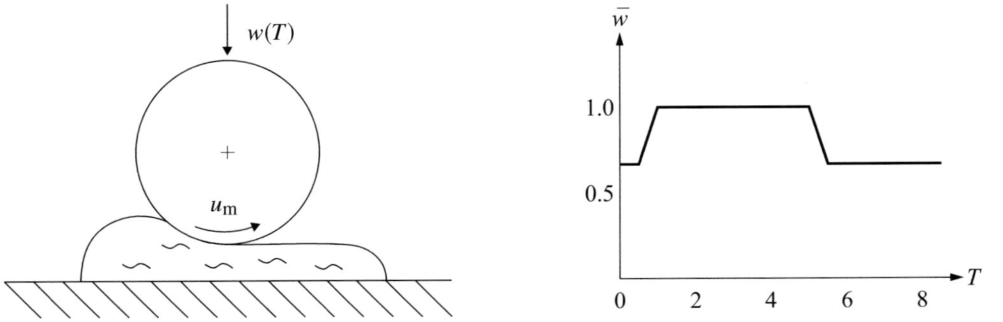


**(a)** Schematic setup of the EHD line contact under combined entraining and squeeze motion.

**(b)** The loading path is given by the nondimensional load $\bar{w}$ over time $T$.

**Figure 4.1:** Schematic setup of the shock loading and unloading problem. Initially, the lubricated contact operates in steady-state conditions. Then the load is rapidly increased and, after some time, rapidly decreased.

### 4.1.1 Example problem

Initially, the lubricated contact operates in steady-state conditions. At nondimensional time $T_1 = 0.5$ the load increases rapidly. Thereafter, the load is kept constant to allow the contact to reach steady-state conditions again. At $T_2 = 5$ the load decreases rapidly.

Figure 4.1b shows the loading path. The loading path is defined by

$$\bar{w}(T) = \begin{cases} \frac{2}{3} & \text{for} \quad T < 0.5, \\ \frac{2}{3} + \frac{1}{3} \cdot \frac{T-0.5}{0.5} & \text{for} \quad 0.5 \leq T \leq 1, \\ 1 & \text{for} \quad 1 < T < 5, \\ 1 - \frac{1}{3} \cdot \frac{T-0.5}{0.5} & \text{for} \quad 5 \leq T \leq 5.5, \\ \frac{2}{3} & \text{for} \quad 5.5 < T. \end{cases} \tag{4.1}$$

The dimensional load is computed as $w(T) = w_{\text{ref}} \cdot \bar{w}(T)$.

**Table 4.1:** Parameters of the shock loading problem.

| parameter | $R$ | $u_{\mathrm{m}}$ | $w_{\mathrm{ref}}$ | $E'$ | $\eta_0$ | $\alpha$ |
|---|---|---|---|---|---|---|
| unit | mm | m/s | N/mm | GPa | Pa·s | 1/GPa |
| value | 22.5 | 5.0365 | 546.11 | 287.8 | 0.01 | 22 |

The conventional EHD line contact model is used, see Appendix B. The lubricant is modeled as compressible, see Eq. (2.10), and the viscosity variation is approximated using Barus' model, see Eq. (2.8). Table 4.1 summarizes the problem's parameters. Figure 4.1 shows the setup of the example problem. The computational domain is defined as $-4 \leq X \leq 2$ with the equidistant grid length $\Delta X = 0.006$. The size of the domain is chosen such that it can adequately consider all relevant effects.

Figure 4.2 shows the spatially resolved pressure (thick line) and gap height (thin line) at different time instances. Figure 4.3 shows the temporal evolution of the central pressure $p_{\mathrm{c}} = p\,(t,\ X \approx 0)$ and central gap height $h_{\mathrm{c}} = h\,(t,\ X \approx 0)$. The upper row of plots in Fig. 4.2 shows the loading phase while the lower row shows the unloading phase.

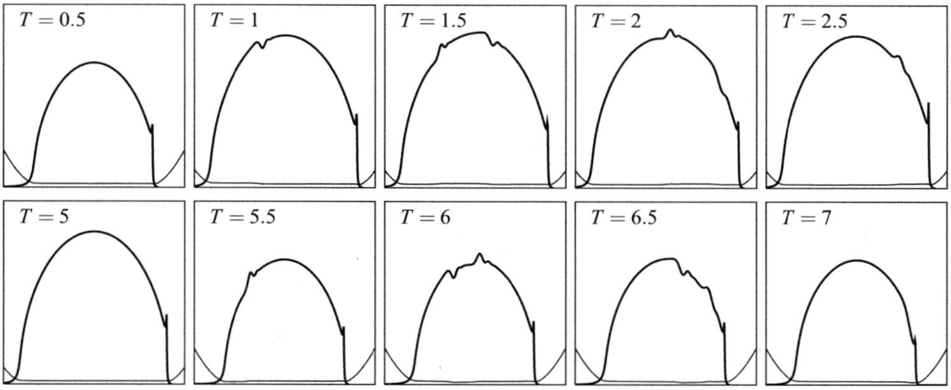


**Figure 4.2:** The plots show the pressure (thick line) and the gap height (thin line) spatially resolved at different time instances. The upper row shows the loading phase and the lower row shows the unloading phase.

As the external load increases, the pressure increases in order to balance the load. Since the viscosity in the high pressure zone is very high, the gap height hardly decreases in the contact's center, but it decreases in the outer contact region, where the viscosity is lower. The squeeze term generates pressure in the outer contact region and thereby hinders the reduction of the gap height. Therefore, for a short time, the gap height entering the contact is larger than it would be in steady-state conditions. The resulting local variation in the gap height and the local pressure peak are transported through the contact region by the entraining motion, as the Poiseuille term in Reynolds Equation vanishes in the high pressure zone. Since no inertial effects are considered, no vibrations occur and the contact reaches steady-state conditions once these local fluctuations have left the contact area after approximately two nondimensional time units. These effects can be observed in Figs. 4.2 and 4.3, furthermore, Wijnant [134] discusses them in further detail.

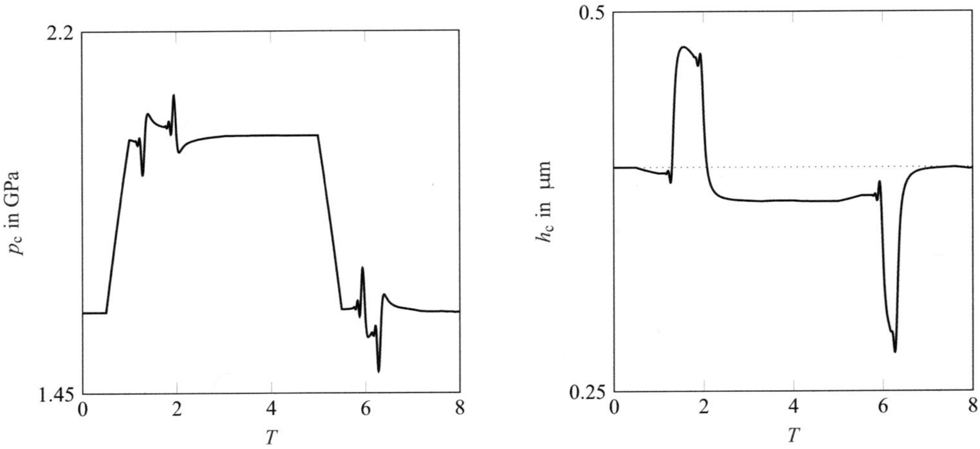


**(a)** **(b)** The dotted line shows the initial gap height.

**Figure 4.3:** Central pressure $p_c$ (a) and central gap height $h_c$ (b) over nondimensional time $T$. The changes in the pressure and the gap height shortly after the load change are particularly visible in this representation.

### 4.1.2 Comparison of time integration schemes

The Backward Euler method and the second-order finite difference method are commonly used in EHD contact problems. This work proposes the use of SDIRK methods, specifically Alexander's second-order method. This subsection compares the three time integration methods.

In the following, the example problem is solved with each method using the fixed dimensionless time step size $\Delta T = 0.01$. The reference solution is computed with second-order finite difference method and the fixed dimensionless step size $\Delta T = 1 \cdot 10^{-4}$. The results of the different methods are compared with a reference solution. Thereafter, the convergence behavior is examined.

Figure 4.4 shows the trace of the central gap height $h_c$ and the minimum gap height $h_{min}(t) = \min_{X \in \Omega_F} (h(t, X))$ for the unloading and loading phase. The Backward Euler method (BE1) shows significant deviations from the reference solution. It is not able to capture the dynamics of the problem and the induced damping is visible. For the second-order finite difference method (FD2), a deviation from the reference solution can be observed as well. However, the dynamics are captured more adequately. The solution computed with Alexander's second-order method (A2) lies close to the reference solution, although it only requires a hundredth of the number of steps. This comparison shows a superior behavior for A2.

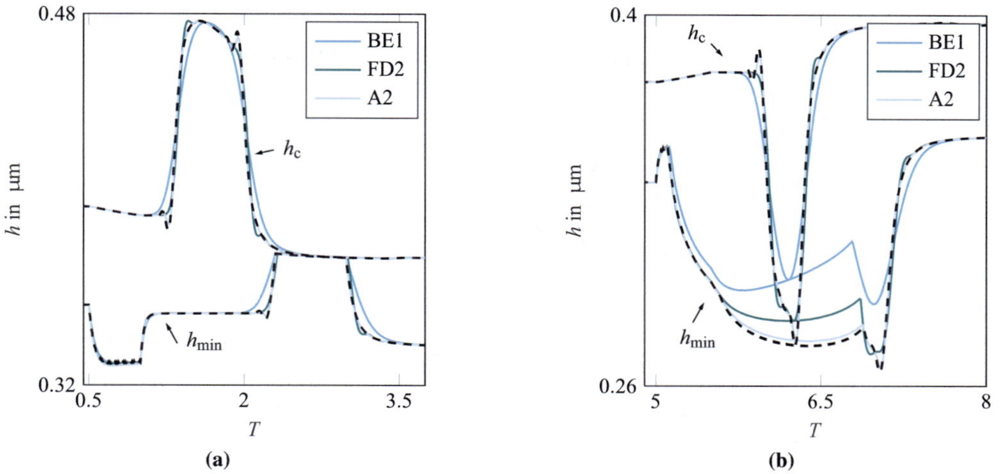


**Figure 4.4:** Central gap height $h_c$ and minimum gap height $h_{min}$ during the loading phase (a) and unloading phase (b) over nondimensional time $T$. The results are computed using Backward Euler method (BE1), second-order finite difference method (FD2) and Alexander's second-order method (A2) with the nondimensional step size $\Delta T = 0.01$. The dashed line is the reference solution. It is computed using FD2 with $\Delta T = 1 \cdot 10^{-4}$.

A study of the convergence order gives further insights into the behavior of a time integration method. It gives information about the functioning of the method for a given problem. This information can be helpful in the context of adaptive time step selection since the used algorithm is based on the convergence order of the method, see Subsection 2.4.2. Additionally, it gives information about the approximate size of the deviation from a reference solution for different time step sizes. This information can be helpful in selecting a suitable time step size for a particular problem.

To carry out an order investigation, the problem is solved with different time step sizes and the deviation from a reference solution is computed. Here, the backward Euler method (BE1), the second-order finite difference method (FD2), and Alexander's second-order method (A2) are investigated. The problem is solved with fixed dimensionless time step sizes $\Delta T = \{0.01, 0.008, 0.006, 0.004, 0.002, 0.001\}$. Since no analytical solution exists for the example problem, the reference solution is computed with FD2 and fixed dimensionless time step size $\Delta T = 1 \cdot 10^{-4}$. The deviation $err_{max}$ of the studied solution from the reference solution is computed using Eq. (A.3).

Figure 4.5 shows the results of the convergence order study for the central pressure $p_c$ and central gap height $h_c$. The Backward Euler method (BE1) does not reach the first order. The reduced order results from its step size dependent numerical damping. Hence, BE1 is not suitable for highly dynamic EHD contact problems. The second-order finite difference method (FD2) reaches second order for smaller step sizes. However, FD2 shows an order reduction for larger step sizes. Alexander's second-order method (A2) yields good results, especially for larger step sizes. For smaller time step sizes, the convergence order decreases slightly. Due to the superior behavior of A2 for larger time step sizes, the deviation from the solution is smaller for A2 than for FD2 in the studied range of time step sizes. The results of the order investigation are in agreement with the results of Fig. 4.4. For more information regarding order reductions, the reader might refer to [99, 105].

In the studies above, Alexander's second-order method shows superior results compared to the other methods. These results show the suitability of the method for dynamic EHD contact problems. Hence, the method is used in the following for the time integration of the Reynolds Equation.

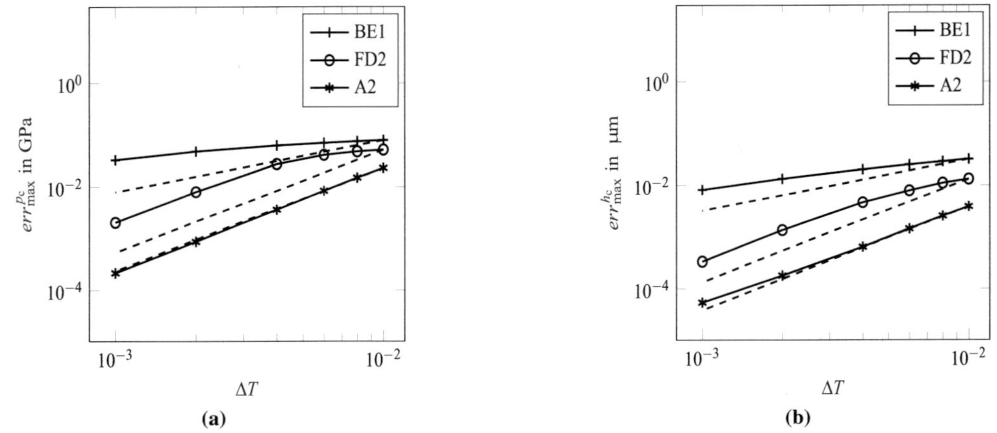


**Figure 4.5:** Order investigation for the central pressure $p_c$ (a) and for the central gap height $h_c$ (b). The slope of the dashed line indicates the theoretical convergence order of the respective method.

### 4.1.3 Study of the adaptive time stepping scheme

To ensure computational stability and efficiency as well as a good approximation of the solution, different time step sizes $\Delta t_j$ must be chosen for different problems. When using a fixed step size, the step size must be selected taking into account the highest possible dynamics of the problem. The resulting small step size may be only necessary for a small part of the problem and, for the most part, a significantly larger step size would be sufficient. As a result, the computational time increases unnecessarily. To this end, a time step size control algorithm should be used to automatically adapt the time step size throughout the temporal path of the problem.

The first part of this subsection aims to compare the adaptive step size control algorithm from Subsection 2.4.2 with procedures from EHD literature. In the second part of this subsection, a procedure to cope with rapid changes of boundary conditions and parameter functions is presented.

**Step size control**

Time step size control algorithms rely on an indicator which is used to approximate a suitable time step size. Commonly, an approximation of the local integration error of the time integration scheme serves as an indicator. One possibility is the approximation based on a linear extrapolation of the previous time step's solutions. This approach has been used in the field of lubricated contacts by [94] and [47]. Alternatively, for SDIRK methods, it is possible to approximate the local integration error based on embedded methods efficiently. This approach is outlined in Subsection 2.4.2. The following study compares the two approaches. Therefore, the following two variants are introduced:

- *Variant FD2-CA1* corresponds to the time integration approach used in some EHD contact problem solvers, such as [94]. Time integration is carried out using second-order finite difference method (FD2). The local integration error is approximated by comparing the previous time step's linear extrapolation to the current time step's result. The approximated error is then used within the time step size control approach CA1, given by Eq. (2.79). This control approach is an adaptation of the algorithm used by [47] and was used in the context of EHD contact computations by [94]. The basic idea of this control approach is to keep the time step size constant while the approximated local integration error is within a predefined range and to change the time step size according to a constant factor, as soon as the approximated error lies outside the range.
- *Variant A2-CA2* employs Alexander's second-order method (A2) and approximates the local integration error based on an embedded scheme, see Subsection 2.4.2, and uses time step size control approach CA2, given by Eq. (2.80). The time step size control approach is an adaptation of the "standard step size controller", see [62, p.124]. It aims to keep the integration error as close as possible to the prescribed tolerance and adjusts the time step size in each time step accordingly.

**Table 4.2:** Overview of the cases studied and their results. The first two cases aim to obtain an approximate relative deviation of $relerr_{\max}^{p_c} \approx 3$ % and the third and fourth case aim to obtain an approximate relative deviation of $relerr_{\max}^{p_c} \approx 1$ %. The absolute tolerance $\epsilon_a$ for adaptive time integration is adjusted accordingly. The other time integration parameters are kept constant, i.e., $\epsilon_r = 0$, $f_{\min} = 0.5$, $f_{\max} = 1.5$ and $f_{\text{safety}} = 0.6$. The relative deviation $relerr_{\max}^{p_c}$ is computed with Eq. (A.5). It gives the deviation from the reference solution, which is computed with the fixed dimensionless time step size $\Delta T_0 = 1 \cdot 10^{-4}$ and FD2 time integration.

| case ID | variant | $\epsilon_a$ | steps (rejected) | $relerr_{\max}^{p_c}$ | rel. comp. time |
|---|---|---|---|---|---|
| FD2-CA1-3% | FD2-CA1 | $2.5 \cdot 10^{-3}$ | 543 (12) | 2.5 % | 1 |
| A2-CA2-3% | A2-CA2 | $5 \cdot 10^{-4}$ | 258 (14) | 2.8 % | 0.947 |
| FD2-CA1-1% | FD2-CA1 | $2.5 \cdot 10^{-4}$ | 2064 (37) | 1.1 % | 1 |
| A2-CA2-1% | A2-CA2 | $5 \cdot 10^{-5}$ | 786 (75) | 0.6 % | 0.828 |

Table 4.2 compares both variants concerning the time steps needed and rejected, relative deviation from a reference solution and relative computational time.[1] The reference solution is computed with fixed dimensionless time step size $\Delta T_0 = 1 \cdot 10^{-4}$ and FD2 time integration. Two different target accuracies are studied. Case 1 and 2 aim to obtain an approximate relative deviation of $relerr_{\max}^{p_c} \approx 3$ % and case 3 and 4 aim for an approximate relative deviation of $relerr_{\max}^{p_c} \approx 1$ %.

A comparison of the number of time steps required shows that FD2-CA1 requires significantly more time steps than A2-CA2 in order to achieve the same accuracy. This is coherent with the observations in the previous subsection, where A2 yields smaller errors than FD2 for the same time step size, see Fig. 4.5.

Comparing the relative computational time of the variants shows that the reduction of time steps needed does not result in a proportionally substantial reduction in computational time. This effect can be explained by the stage values, which need to be computed in higher order implicit Runge-Kutta methods for each time step, see Section 2.4. The computation of stage values requires recomputations of the Jacobian matrix, see Eq. (3.54), and evaluations of the nonlinear system, see Eq. (3.48).

Figure 4.6 shows the time step size $\Delta T_j$ over time $T$ for both target accuracies. It shows that step size control works for both variants, as the time step size in the stationary phases is increased and the load changes in the environment are significantly reduced. Furthermore, this figure exemplifies the differences of the two control approaches CA1 and CA2, given by Eq. (2.79) and Eq. (2.80), respectively.

The control approach CA1 keeps the time step size constant while the approximated local integration error is within a predefined range. If the approximated error leaves that range, the

[1] There is no direct relation between the prescribed tolerance $\epsilon_a$ and the relative deviation $relerr_{\max}^{p}$. The tolerance $\epsilon_a$ is compared to the root mean square of all field variables, see Eq. (2.78), whereas $relerr_{\max}^{p}$ gives the relative deviation of the central pressure $p_c$ according to Eq. (A.5).

time step size is adjusted according to a constant factor. This approach leads to a step-wise path of the time step size and minimizes the number of time step size changes.

The control approach CA2 adjusts the time step size such that the approximated local integration error is as close as possible to the predefined tolerance. This procedure gives a smooth path of the time step size and yields a large number of time step size changes. The increased number of time step size changes also increases the likelihood of a rejected time step, as can be seen in Tab. 4.2.

Rejected time steps are expensive because a solution of the whole nonlinear system must be computed and then discarded. In this example problem, long steady state phases are followed by rapid load changes. The rapid load changes require a rapid decrease of the time step size, which leads to rejected time steps in both variants. In the following, a procedure of how to deal with rapid load changes is introduced.

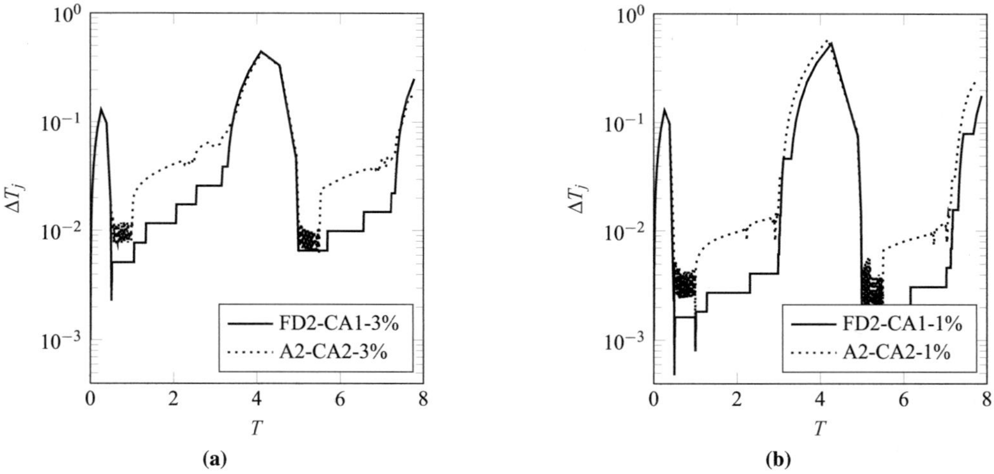


**Figure 4.6:** Time step size $\Delta T_j$ over time $T$ for desired approximate relative deviation of $relerr_{\max}^{p_c} \approx 3$ % (a) and $relerr_{\max}^{p_c} \approx 1$ % (b). Table 4.2 summarizes the characteristics of the different cases.

### Considering the integration error of prescribed functions in time step size selection

The example problem consists of phases with constant load and highly dynamic phases in which the load changes rapidly. In order to take these very rapid load changes and the subsequent dynamics into account, the step size must be reduced rapidly. However, rapid changes in boundary conditions and parameters might not be efficiently captured by a standard step size control algorithms. Hence, the rapid reduction of the step size may cause several rejected step sizes. To this end, an extension to the time-stepping algorithm from Subsection 2.4.2 is proposed here.

With the extensions, the local integration error of predefined functions describing boundary conditions and parameters is considered for the selection of the time step size.

Figure 4.6 shows a substantial increase of the time step size $\Delta T_j$ starting from $T \approx 3$ until right before the rapid load decrease at $T = 5$. During this steady-state phase, the field variables are constant and their integration error vanishes. Consequently, the time step size increases. The time step size becomes so large that it skips over the unloading phase from $T = 5$ to $T = 5.5$. This causes the solution procedure to diverge and as a consequence, the time step is rejected. A new computation with a smaller time step size will likely be discarded again, since very small time step sizes are needed to capture the rapid change in the prescribed load properly. The time step size will be reduced iteratively. For each attempt, several Newton-Raphson iterations of the full system are carried out, which results in computational work.

In an attempt to reduce the number of rejected steps during the rapid change of the prescribed load, more advanced approaches based on control theory, as given in [53, 116], are tried. They do not yield a significant improvement for the studied problem and are not further investigated within the scope of this work. However, a procedure to reduce the number of rejected time steps is proposed in the following.

The procedure adjusts the time step size such that the local integration error of the predefined load function stays below a certain threshold $\epsilon_w$. This procedure can be applied to any predefined functions, e.g., parameter or boundary condition functions. The flowchart in Fig. 4.7 outlines this procedure. It ensures that time-dependent boundary conditions and parameters are considered by the adaptive step size selection method. Thereby, it reduces the risk that time steps are rejected due to rapid changes of a predefined function. It allows for using larger time integration tolerances throughout the complete simulation while making sure rapid changes in boundary conditions are still considered in detail. The advantage of this approach compared to the conventional approach, i.e., to consider only the field variables, is that the evaluation of a predefined function generally requires considerably less effort than the computation of several iterations of the complete system.

To study the procedure described above, the same cases as above are recomputed, see Tab. 4.2. However, the local integration error of the load function is considered in step size selection using the procedure above with a predefined threshold of $\epsilon_w = 1 \cdot 10^{-3}$.

Table 4.3 summarizes the results. In all cases, the number of rejected steps and the computational time decreases. The reduction is specially useful for higher tolerances $\epsilon_a$, see cases 1 and 2. Due to the larger tolerances, the overall time step size is larger. The larger time step size increases the risk of skipping over rapid parameter changes, which leads to rejected time steps. For case 3 and 4 this effect is less pronounced. However, the procedure still yields a reduction in computational time. The procedure yields better results for variant A2-CA2.

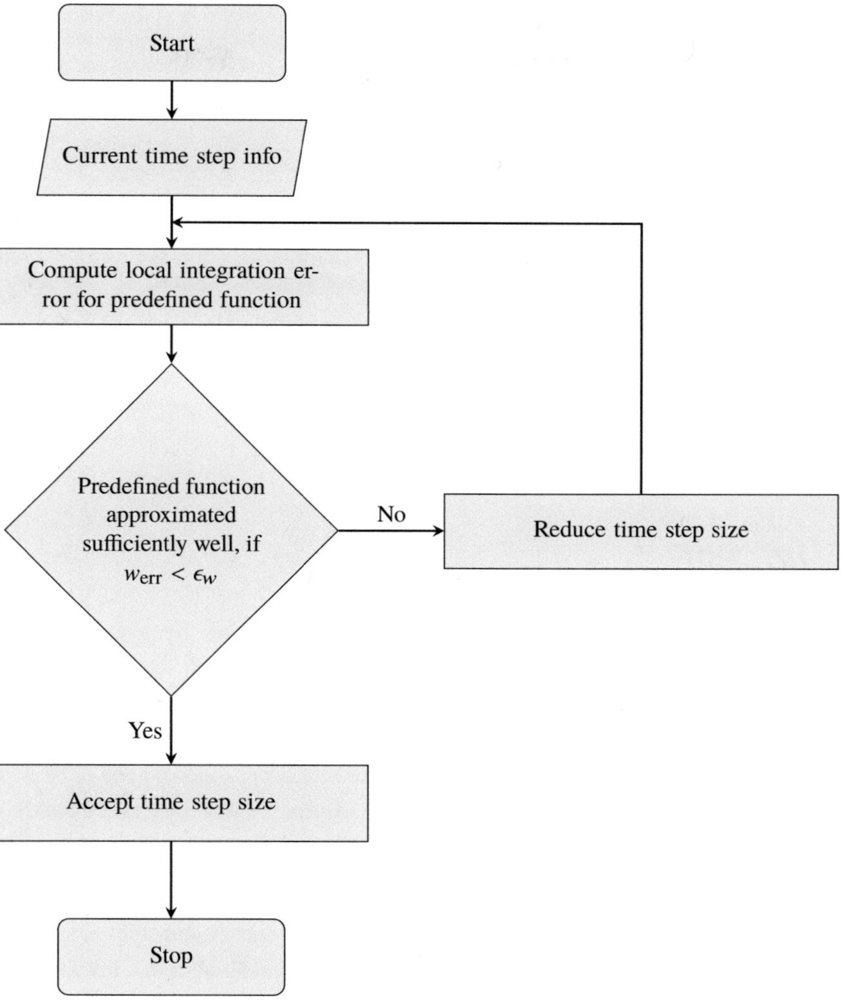


**Figure 4.7:** This procedure considers the local integration error for a predefined function during the selection of the time step size. In this problem, the predefined function describes the temporal evolution of the load, see Eq. (4.1). For each time step, the local integration error of the function is computed. The aim is to reduce the number of rejected time steps and thereby to reduce the computational effort to solve the problem. Hence, the computation of the local integration error should not be computationally costly. Appendix C outlines the procedure to compute the local integration error. The following steps describe the general procedure to approximate the local integration error: 1. Compute the temporal derivative of the function using complex step differentiation. 2. Use numerical time integration to compute the approximate function value. 3. Compute the deviation of the approximated function value from the exact function value. This comparison gives the approximation of the local integration error. If the error is above the prescribed threshold $\epsilon_w$, the time step size is successively reduced until the error is below the threshold.

**Table 4.3:** Overview of results with the consideration of the local integration error of the load function $w(T)$. All cases are computed with the same parameters as in Tab. 4.2. The threshold for the local integration error of the load function is chosen to be $\epsilon_w = 1 \cdot 10^{-3}$. The relative computational time is computed relative to the respective case without monitoring the load function. As above, the reference solution is computed with fixed dimensionless time step size $\Delta T_0 = 1 \cdot 10^{-4}$ and FD2 time integration.

| case ID | variant | $\epsilon_a$ | steps (rejected) | $relerr^p_{max}$ | rel. comp. time |
|---|---|---|---|---|---|
| FD2-CA1-3% | FD2-CA1 | $2.5 \cdot 10^{-3}$ | 561 (0) | 2.7 % | 0.870 |
| A2-CA2-3% | A2-CA2 | $5 \cdot 10^{-4}$ | 263 (3) | 2.8 % | 0.877 |
| FD2-CA1-1% | FD2-CA1 | $2.5 \cdot 10^{-4}$ | 2112 (36) | 1.5 % | 0.975 |
| A2-CA2-1% | A2-CA2 | $5 \cdot 10^{-5}$ | 781 (58) | 0.6 % | 0.933 |

The results indicate that Alexander's second-order method with the control approach given by Eq. (2.80) and monitoring of prescribed functions yields good results in EHD contact problems. Thus, this method is used for time integration in the following.

## 4.2 Dry contact

Conventional EHD contact models consider the solid's dynamic behavior in terms of rigid body motion and its deformation via an elastic half-space, see [128, 133, 134]. This modeling approach is limited as it does not account for the structural dynamic behavior of complex geometries. This work proposes to extend the structural dynamics model of EHD contacts, using a reduced order modeling approach. However, the reduced order models need to be adjusted in a specific manner in order to yield accurate and stable results. This section examines the proposed procedure and discusses the model's suitability for dynamic contact problems.

While the previous section focused on the fluid part of the coupled problem, this section focuses on the structural dynamics part. To isolate the structural dynamics component, other physical effects, such as the behavior of the lubricant, shall not be considered in this investigation. Hence, a dry contact problem is examined. The low-velocity impact of a plane T-shaped elastic body on a rigid plate is used as an example problem, see Fig. 4.8. The shape of the upper body is motivated by the shape of valve needles as they can be found in injectors. The geometry and the mass distribution of this body introduce structural dynamic effects, which shall be captured by the extended structural model.

The first subsection describes the example problem. Subsection 4.2.2 discusses the construction of a reduced order model suitable for the problem using the concepts introduced in Section 2.3. The results of this subsection show the occurrence of spurious oscillations. Subsection 4.2.3 examines two different approaches to minimize these oscillations. At the end of this subsection, a suitable structural dynamic model is given. While the first subsections focus on the construction of the structural dynamic model, Subsection 4.2.4 examines the proposed time integration procedure for the structural dynamic part of the coupled problem.

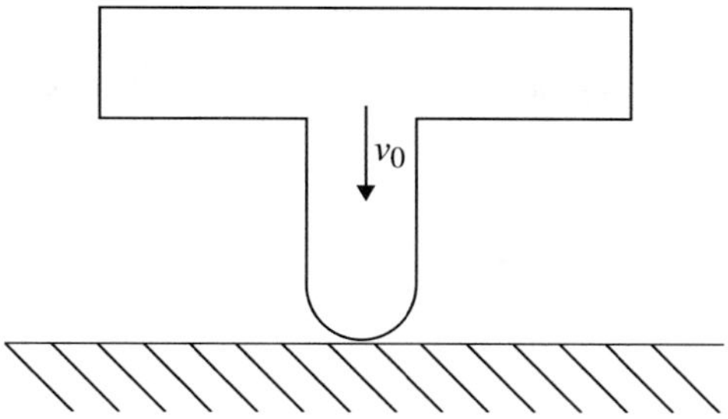


**Figure 4.8:** Schematic of the example problem setup. The elastic T-geometry moves with the initial velocity $v_0$ towards the rigid plate. It impacts the plate and rebounds from it. After the rebound it impacts the plate again because of the geometrical shape of the body and its elastic deformation.

### 4.2.1 Example problem

Figure 4.8 shows the example problem setup. The upper body moves with the prescribed initial velocity $v_0 = 10$ m/s towards the rigid plate. The initial displacement is zero, i.e., the undeformed upper body is touching the lower body.

The T-geometry is modeled to behave linear-elastically under plane-strain condition with $E = 100$ GPa and $\nu = 0.3$. It's mass is $m_0 = 5.513$ kg/m. The body is discretized using finite elements with quadratic shape functions. The lower body is modeled as a rigid plate.

The contact between the bodies is assumed frictionless. The contact conditions are enforced by the penalty contact formulation

$$f_{p,i} = P - \gamma_{\text{pen}} \min(H_i, 0)^2 = 0, \tag{4.2}$$

with the penalty factor $\gamma_{\text{pen}}$.[2] Using this expression, the dimensionless contact pressure $P_i$ is found such that the dimensionless gap height $H_i$ at each node $i = 1, 2, \ldots, n_{\text{n}}^{\text{u}}$ does not become negative. It was found that using the squared gap height in Eq. (4.2) improves the convergence of the Newton-Raphson method, because this way $H_i$ occurs in the derivative of $f_{p,i}$ with respect to $H_i$ and thus increases the information content of the Jacobian matrix. Evaluating Eq. (4.2) for all contact nodes gives the vector $\boldsymbol{f}_p \in \mathbb{R}^{n_{\text{n}}^{\text{u}}}$, which is inserted instead of the Reynolds Equation into the nonlinear system (3.48). Section 3.6 describes the solution procedure of the resulting nonlinear system.

[2]The penalty parameter should be chosen as high as possible to enforce contact conditions properly. However, high penalty parameters might yield numerical problems. In such cases, it should be reduced. In this section, the penalty parameter is chosen to be $\gamma_{\text{pen}} = 1 \cdot 10^{10}$.

For comparison and validation purposes, reference solutions are computed with the commercial finite element software Abaqus. In Subsections 4.2.2 and 4.2.3 these results are computed using Hilber-Hughes-Taylor time integration method with the parameters $\alpha = -0.05, \beta = 0.275625$, and $\gamma = 0.55$. These parameters are recommended for elastodynamic contact problems by [29, 30]. The method with these parameters will be referred to as HHT-TF in the following.

### 4.2.2 Construction of the structural reduced order model

The structural dynamics model is given as a reduced order model. The reduced order model is constructed using component mode synthesis. Depending on the application, different properties of a reduced order model are relevant. The properties of a reduced order model can be significantly influenced by three factors: the reduction method, the cut-off frequency $f_c$, and the number and location of the master nodes. These factors should be chosen such that the resulting reduced order model is well suited for the application at hand. The following study examines the influence of these factors so they can be selected appropriately.

Subsection 2.3.5 introduces two component mode synthesis methods: Craig-Bampton method and an adapted Craig-Chang method. These two methods, as well as the cut-off frequency $f_c$ and the number and location of the master nodes, will be compared in the following. To this end, six different models are created. Three models are created for each reduction method. The three models vary in cut-off frequency and master node configurations. The master node configurations are shown in Fig. 4.9. In the CN-configuration, see Fig. 4.9a, the nodes in the contact region are selected as master nodes. In the AN-configuration, see Fig. 4.9b, additional nodes outside of the contact region are selected. Table 4.4 gives an overview of the reduced order models.

**Table 4.4:** Overview of the studied reduced order models. The models are referred to by their model ID in the following.

| model ID | reduction method | cut-off frequency $f_c$ | master node configuration |
|---|---|---|---|
| CC-150-CN | Craig-Chang method | 150 kHz | as in Fig. 4.9a |
| CC-150-AN | Craig-Chang method | 150 kHz | as in Fig. 4.9b |
| CC-450-CN | Craig-Chang method | 450 kHz | as in Fig. 4.9a |
| CB-150-CN | Craig-Bampton method | 150 kHz | as in Fig. 4.9a |
| CB-150-AN | Craig-Bampton method | 150 kHz | as in Fig. 4.9b |
| CB-450-CN | Craig-Bampton method | 450 kHz | as in Fig. 4.9a |

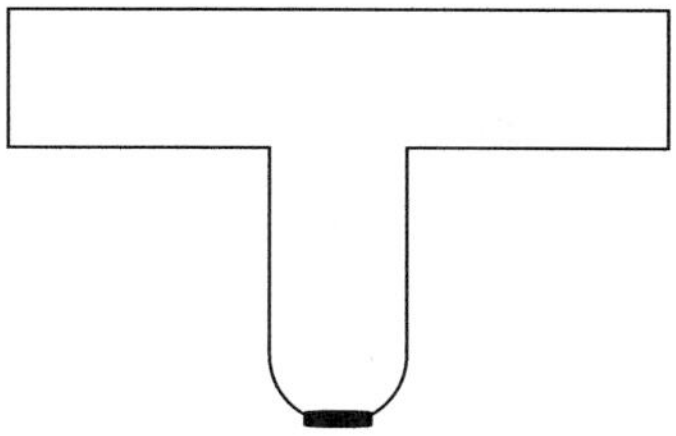

**(a)** Only the nodes in the contact region (highlighted black) are selected as master nodes. The corresponding reduced order model is referred to as CN, short for contact nodes.

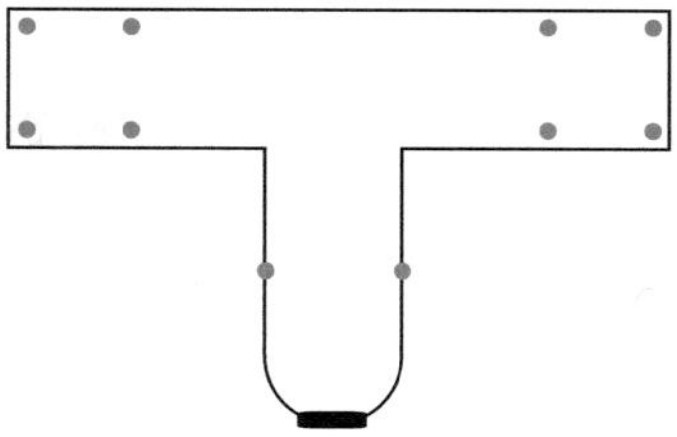

**(b)** Besides to the nodes in the contact region (highlighted black) additional nodes (highlighted gray) outside of the contact region are selected as master nodes. The corresponding reduced order model is referred to as AN, short for additional nodes.

**Figure 4.9:** The highlighted nodes represent the master nodes selected in the CN-configuration (a) and the AN-configuration (b).

A corresponding reference solution is computed with the commercial finite element software Abaqus using HHT-TF time integration with fixed time step size $\Delta t = 6\,\text{ns}$. The reference solution is referred to as the *full model*. The solutions of the reduced order models are computed with the Matlab implementation using Alexander's second-order method (A2) and fixed time step size $\Delta t = 12\,\text{ns}$.

Figure 4.10a shows the central force $F_c = F(t, X \approx 0)$, which is the force at the node closest to the contact's center, and the central gap height $h_\text{c}$ over time $t$ for the Craig-Chang method based reduced order models. The solution of the full model is also shown as a reference. Model CC-150-CN uses the CN-configuration as master nodes and the cut-off frequency is set as $f_\text{c} = 150$ kHz. It yields strong contact force oscillations. The force oscillations are so strong that the central node briefly leaves the contact at around 17 µs. Increasing the cut-off frequency from $f_\text{c} = 150$ kHz to $f_\text{c} = 450$ kHz, as done for model CC-450-CN, reduces the oscillations significantly and increases the overall accordance with the reference solution. However, the contact time remains underestimated and the rebound height remains overestimated. For model CC-150-AN additional master nodes are selected as per the AN-configuration and the cut-off frequency is set to $f_\text{c} = 150$ kHz. This model shows the best results of the three studied variants. However, for all models constructed with Craig-Chang method the contact time is underestimated and the rebound height is overestimated. All reduced order models studied show spurious oscillations.

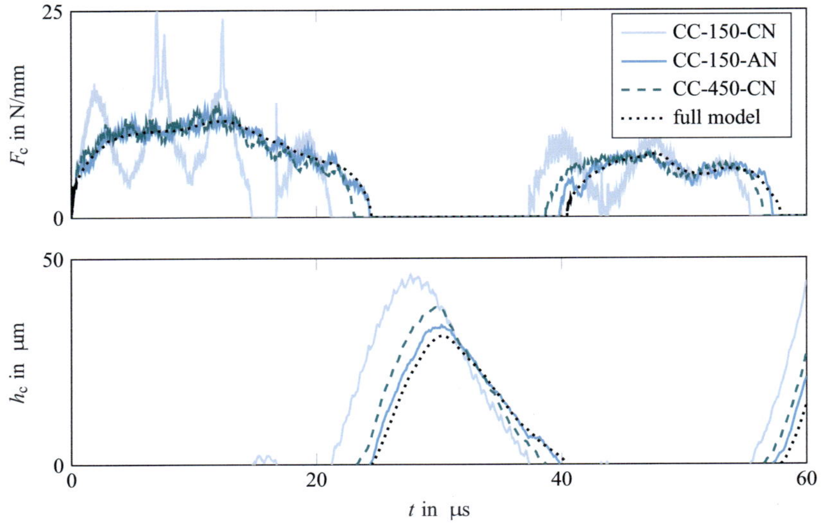


**(a)**

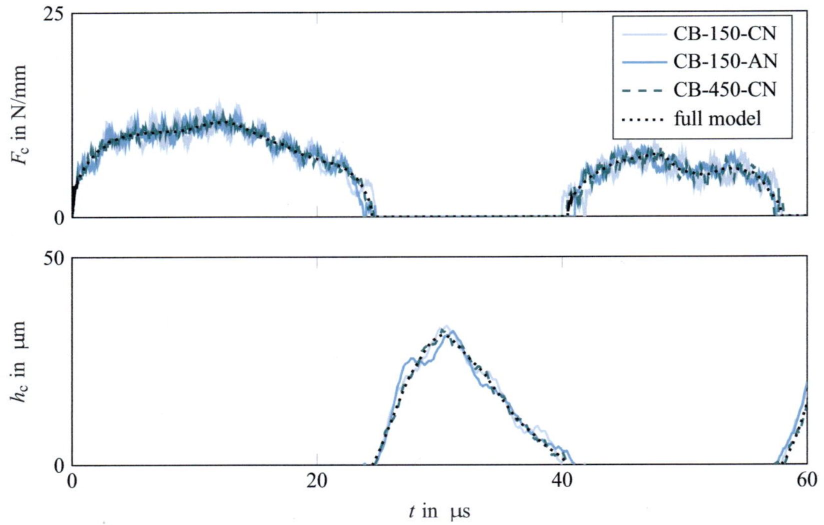


**(b)**

**Figure 4.10:** Central force $F_c$ (upper subfigure) and central gap height $h_c$ (lower subfigure) over time $t$. The results are computed with Craig-Chang method based reduced order models (a) and Craig-Bampton method based reduced order models (b). The results of the full model are plotted for comparison. The characteristics of the reduced order models are summarized in Tab. 4.4.

The Craig-Bampton method based models behave differently. Figure 4.10b shows the results of reduced order models constructed using Craig-Bampton method. The contact force oscillations of CB-150-CN are much less pronounced as compared to the respective Craig-Chang method based model, see CC-150-CN in Fig. 4.10a. Increasing the cut-off frequency from $f_c$ = 150 kHz to $f_c$ = 450 kHz, as done for model CB-450-CN, reduces the spurious oscillations and improves the overall accordance with the reference solution. Introducing additional master nodes as per the AN-configuration, while keeping the cut-off frequency at $f_c$ = 150 kHz, as done for model CB-150-AN, reduces the accordance of the reduced order model with the full model. The deteriorated accordance can be seen in the gap height during the flight phase. The accordance of the results regarding contact time and rebound height is good for the Craig-Bampton method based models. As for the Craig-Chang method based models, all reduced order models studied show spurious oscillations, which decrease with increasing cut-off frequency.

Above observations show that increasing the cut-off frequency $f_c$ improves the model in both reduction methods. This can be explained by the fact that the increase of the cut-off frequency $f_c$ increases the information content of the normal mode matrix $\mathbf{\Phi}_{\mathrm{nm}}$. The information content of the constraint mode matrix $\mathbf{\Phi}_{\mathrm{cm}}$ remains unchanged. The increased number of normal modes improves the approximation of the model's dynamics and reduces the influence of the artificial frequencies. Independent of the reduction method, increasing the cut-off frequency increases the information content of the model and thereby improves the quality of the reduced order model.

Including additional master nodes increases the information content of the constraint mode matrix $\mathbf{\Phi}_{\mathrm{cm}}$. However, depending on the reduction method, including additional master nodes has different implications on the normal mode matrix $\mathbf{\Phi}_{\mathrm{nm}}$ and on the overall result.

With the Craig-Chang method, the master nodes are unconstrained, i.e., free. Hence, the choice of master nodes does not influence the normal modes. However, it increases the information content of the constraint modes. Hence, for the Craig-Chang method, choosing additional master nodes increases the overall information content of the reduced order model and thereby increases its quality.

With the Craig-Bampton method the master nodes are fixed during normal mode extraction. Therefore, the choice of master nodes influences the normal modes. For the given example problem, additional master nodes are located on the horizontal bar of the T-geometry. These additional nodes limit the dynamics of the geometry during modal analysis. As a result, less information about the geometry's dynamic behavior is captured by the normal modes. The additional master nodes increase the information content of the constraint modes. However, for the given example problem, the increased dynamics information content of the constraint modes does not make up for the reduced dynamics information content of the normal modes. Hence, the accordance of the results decreases for this example problem.

The choice of master nodes must be carried out with greater care for the Craig-Bampton method. It must be considered what mode shapes are expected and what master nodes will be constrained or will come into contact during the application of the reduced order model. Hence, additional knowledge of the problem is needed. Taking this into account, improved reduced order models could be created with both methods by considering additional master nodes. Also, hybrid methods could be developed where only a part of the master nodes is fixed for modal analysis. However, these studies are out of the scope of this work.

The influence of an increased cut-off frequency or additional master nodes can also be seen in the eigenfrequencies of the reduced order model. Figure 4.11 shows the eigenfrequencies of the full model and the reduced order models. Figure 4.11a shows the results for Craig-Chang method whereas Fig. 4.11b shows the results for Craig-Bampton method. Fewer differences between the two reduction methods are observable concerning the eigenfrequencies. The eigenfrequencies of the reduced order models are in good accordance with the full model until close to the respective cut-off frequency. The eigenfrequencies diverge quickly for frequencies larger than the respective cut-off frequency. In the AN-configuration, i.e., when additional master nodes are selected, the eigenfrequencies diverge more slowly than in the CN-configuration. However, in all cases, the accordance of the eigenfrequencies is bad for frequencies above the cut-off frequency.

In summary, the Craig-Bampton method based models yield superior results to the Craig-Chang method based models. Especially, the contact time and the flight phase are better approximated by Craig-Bampton method based models. This behavior can be explained by the better approximation of the contact conditions by the fixed boundary conditions on the contact nodes. Hence, the following studies are carried out with Craig-Bampton method based models.

Independent of the method, all reduced order models show spurious oscillations. The following subsection investigates approaches to reduce these oscillations.

### 4.2.3 Study of artificial frequencies

The previous subsection investigates two different structural reduction methods. It concludes that the Craig-Bampton method based models with master nodes selected according to the CN-configuration shows good results for the given problem and will thus be used in the following. However, spurious oscillations occur independently of the reduction method. The spurious oscillations are introduced into the model by the constraint mode matrix $\mathbf{\Phi}_{\mathrm{cm}}$. The constraint modes are needed to approximate the static displacements in the contact region. The improved modeling in the contact region, however, is accompanied by spurious oscillations in the temporal domain, which are visible in the force curve, see Fig. 4.10. The spurious oscillations are caused by artificial frequencies, which can be seen by the substantial rise of the eigenfrequencies above the cut-off frequency, see Fig. 4.11. In order to obtain good results, the spurious oscillations need to be reduced.

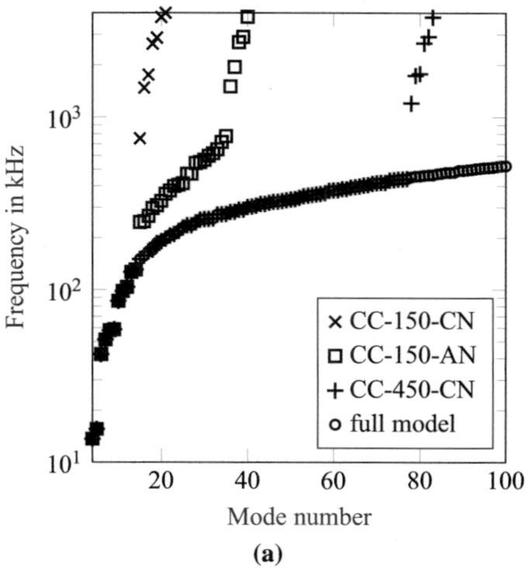


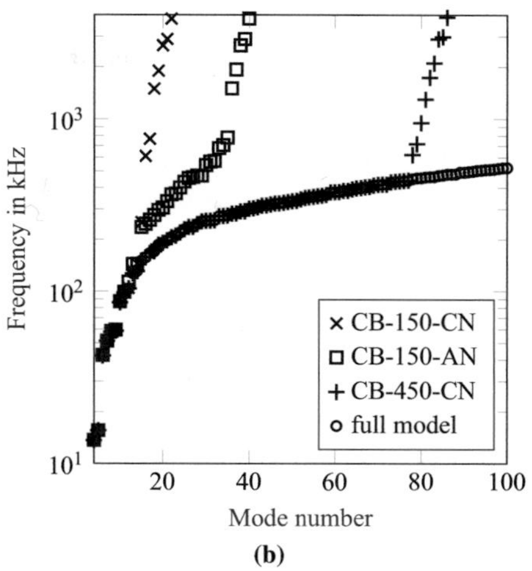


**Figure 4.11:** Eigenfrequencies of the reduced order models constructed with Craig-Chang method (a) and constructed with Craig-Bampton method (b) are compared with the eigenfrequencies of the full model. Until close to the respective cut-off frequency, the eigenfrequencies of all cases are almost identical. The eigenfrequencies of the reduced order models diverge quickly from the eigenfrequencies of the full model for frequencies above the respective cut-off frequencies. From the divergence, the artificial frequencies are visible for both reduction approaches.

Section 3.3 discusses different approaches to reduce spurious oscillations. These approaches are referred to as the *damping approach* and as the *feedthrough approach.*

- In the damping approach, the statically considered modes are damped to reduce their dynamic influence. Here, modal damping is used. In this approach, the number of dynamically considered modes $n_{\mathrm{dc}}$ and the damping ratio $\zeta$ must be specified. These two parameters must be carefully selected as they strongly influence the quality of the solution.
- In the feedthrough approach, the statically considered modes are not taken into account in time integration. Their contribution is only considered statically in the displacement computation, see Eq. (3.11). In this approach, no damping ratio needs to be prescribed. This approach only depends on the number of dynamically considered modes $n_{\mathrm{dc}}$, which strongly influences the quality of the results.

The following study examines the two approaches. In the first part of the study, the influence of the approaches on the quality of the solution is examined. For this purpose, with the damping approach, the number of dynamically considered modes $n_{\mathrm{dc}}$ and the damping ratio $\zeta$ are varied. With the feedthrough approach, only the number of dynamically considered modes $n_{\mathrm{dc}}$ is varied. In the second part of the study, the approaches are investigated concerning computational time. Finally, two models with different cut-off frequencies $f_{\mathrm{c}}$ are compared.

The first part of the study examines eight different cases. Six cases are using the damping approach and two cases are using the feedthrough approach. The number of dynamically considered modes

$n_{dc}$ is varied for both approaches. Since the constraint modes cause the spurious oscillations, it stands to reason not to consider them dynamically. This means the number of dynamically considered modes is given as the number of all modes $n_q$ minus the number of constraint modes $n_{cm}$, i.e., $n_{dc} = n_q - n_{cm}$. For model CB-150-CN, this number is $n_{dc} = 13$, see Tab. 4.4. For the damping approach, the damping ratio $\zeta$ is varied as well. The spurious oscillations are more pronounced for models with lower cut-off frequencies. Hence, in order to make the examined effects more visible, this part of the study uses a model with cut-off frequency $f_c = 150$ kHz, i.e., CB-150-CN. Table 4.5 gives an overview of the cases studied.

The solutions of the reduced order models are computed with the Matlab implementation using Alexander's second-order method (A2) and fixed time step size $\Delta t = 12$ ns. The reference solution is computed with the commercial finite element software Abaqus using HHT-TF time integration with fixed time step size $\Delta t = 6$ ns.

**Table 4.5:** Overview of cases used to study the different approaches to reduce spurious oscillations. The models referred to by the model ID are listed in Tab. 4.4. In the following figures the different cases are referred to by their case ID.

| case ID | approach | $\zeta$ | $n_{dc}$ | $n_{cm}$ | $n_q$ | model ID |
|---|---|---|---|---|---|---|
| D-0.5-13 | damping | 0.5 | 13 | 482 | 495 | CB-150-CN |
| D-0.5-15 | damping | 0.5 | 15 | 482 | 495 | CB-150-CN |
| D-0.5-20 | damping | 0.5 | 20 | 482 | 495 | CB-150-CN |
| D-0.1-115 | damping | 0.1 | 15 | 482 | 495 | CB-150-CN |
| D-0.5-15 | damping | 0.5 | 15 | 482 | 495 | CB-150-CN |
| D-1.0-15 | damping | 1.0 | 15 | 482 | 495 | CB-150-CN |
| FT-13 | feedthrough | N/A | 13 | 482 | 495 | CB-150-CN |
| FT-15 | feedthrough | N/A | 15 | 482 | 495 | CB-150-CN |
| FT-77 | feedthrough | N/A | 77 | 482 | 559 | CB-450-CN |

## Damping approach

Figures 4.12 and 4.13 show the results of the cases using the damping approach. Figure 4.12 shows the central force $F_c$ (a) and central gap height $h_c$ (b) for various numbers of dynamically considered modes $n_{dc}$ and a fixed damping ratio of $\zeta = 0.5$. Likewise, Fig. 4.13 shows the respective results for various damping ratios $\zeta$ and $n_{dc} = 15$. In all cases, the results of the full model are shown for comparison. The relative deviation $relerr$ of the results of the reduced order models from the full model is shown in the lower sub-plot. The relative deviation $relerr$ is computed using Eq. (A.4).

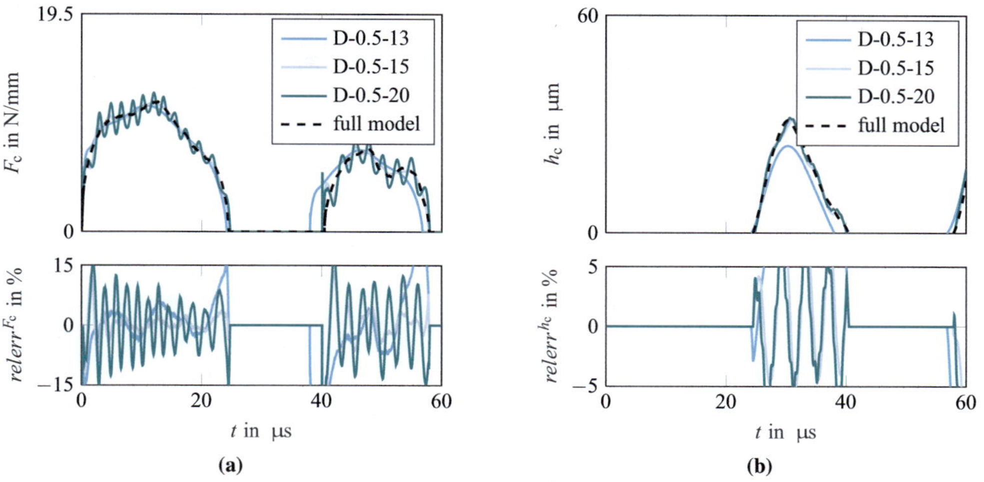


**Figure 4.12:** Results of the damping approach for various numbers of dynamically considered modes $n_{dc}$ and a fixed damping ratio of $\zeta = 0.5$. The upper plots show the central force $F_c$ (a) and central gap height $h_c$ (b) over time $t$ and the lower plots show the respective relative deviation of the reduced order model from the full model $relerr^{F_c}$ (a) and $relerr^{h_c}$ (b) over time $t$.

Varying the number of damped modes and the damping ratio yields the following behavior:

- Figure 4.12 shows the results for the variation of the number of dynamically considered modes $n_{dc}$. For a lower number of dynamically considered modes, see case D-0.5-13, the accordance with the solution of the full model is not good. While the spurious oscillations are reduced, the flight time as well as the rebound height are underestimated. The underestimation of the flight time and rebound height results from a loss of energy due to damping. This observation indicates that too many modes are damped with a too large damping ratio. Increasing the number of dynamically considered modes to $n_{dc} = 15$ improves the approximation of the dynamics of the system, see case D-0.5-15. However, some oscillations in the central force $F_c$ are introduced. These oscillations are damped quickly by the large damping ratio. Increasing the number of considered modes further, see case D-0.5-20, introduces further spurious oscillations and reduces the accordance. With an increasing number of dynamically considered modes, the solution converges towards the untreated solution. For model CB-150-CN, $n_{dc} = 15$ yields the best results.
- Figure 4.13 shows the results for the variation of the damping ratio $\zeta$. A damping ratio of $\zeta = 0.1$ introduces too little damping to reduce the spurious oscillations effectively, see case D-0.1-15. Increasing the damping ratio to $\zeta = 0.5$ reduces the oscillations significantly and improves the accordance with the solution of the full model. A further increase of the damping ratio to $\zeta = 1.0$ yields no significant effect.

Overall, the damping approach significantly decreases the spurious oscillations. The approach yields better results when few additional modes corresponding to the constraint modes are considered dynamically, i.e., $n_{dc} > n_q - n_{cm}$. This observation can be explained by the fact that the artificial modes are relevant to approximate the deformation in the contact zone. If these modes are damped too much, the deformation speed is reduced strongly, which results in nonphysical results, i.e., the observed underestimation of the flight time and the rebound height.

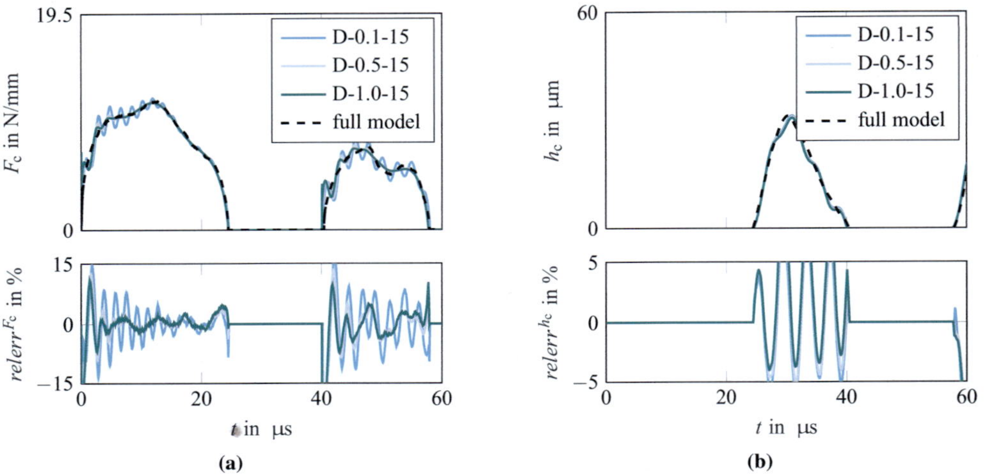


**Figure 4.13:** Results of the damping approach for various damping ratios $\zeta$ and $n_{dc} = 15$. The upper plots show the central force $F_c$ (a) and central gap height $h_c$ (b) over time $t$. The lower plots show the respective relative deviation of the reduced order model from the full model $relerr^{F_c}$ (a) and $relerr^{h_c}$ (b) over time $t$.

### Feedthrough approach

Figure 4.14 shows the results obtained with the feedthrough approach. For case FT-13, good accordance with the full model is observed. The contact time, as well as the rebound height, are well approximated with $n_{dc} = 13$. This observation can be explained by the fact that the artificial modes are not damped in the feedthrough approach. The artificial modes are simply not considered dynamically. However, their static influence is taken into account.

For an increased number of dynamically considered modes, i.e., $n_{dc} = 15$, strong oscillations occur, see FT-15. These oscillations are observable for both the central force $F_c$ and the central gap height $h_c$, see Fig. 4.14. The amplitude of the oscillations remains almost constant over the entire contact time. This behavior is expected because no damping is prescribed to the system. It can be seen that considering artificial modes dynamically without damping leads to strong oscillations.

The feedthrough approach yields better results when no artificial modes are considered dynamically. Therefore, no parameter selection is necessary for the feedthrough approach. The number of dynamically considered modes can be selected according to the formula $n_{dc} = n_q - n_{cm}$.

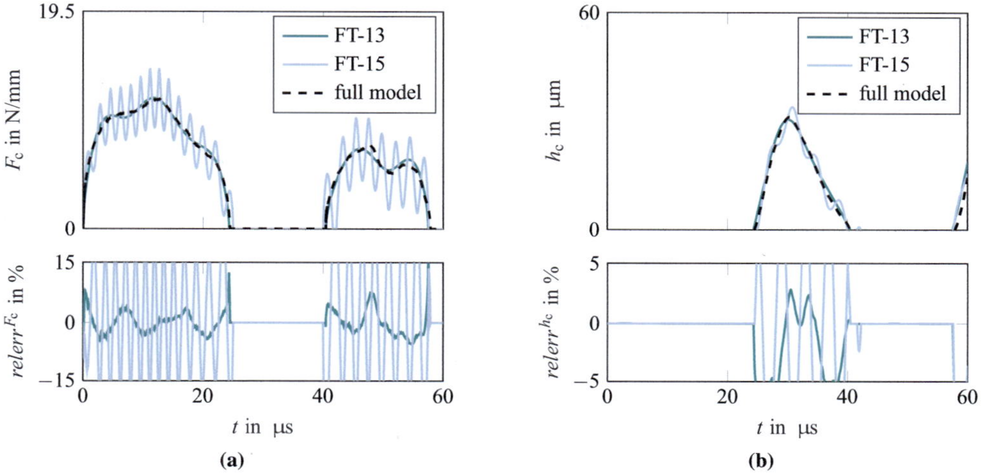


**Figure 4.14:** Results of the feedthrough approach. The upper plots show the central force $F_c$ (a) and central gap height $h_c$ (b) over time $t$. The lower plots show the respective relative deviation of the reduced order model from the full model $relerr^{F_c}$ (a) and $relerr^{h_c}$ (b) over time $t$.

## Computational time

The study above focuses on aspects regarding solution quality. Another advantage of the feedthrough approach results from the reduced number of the state equations involved in time integration, see Eq. (3.11b). In the feedthrough approach, only $n_{dc}$ variables are considered during time integration, which reduces the number of states from $n_q$ to $n_{dc}$. This reduction of states reduces the computational time of the linear structural model. Especially for linear systems, the time savings are significant.[3]

Here, for the nonlinear coupled problem, most of the computational time is used to solve the nonlinear system of equations. Therefore the time savings of the feedthrough approach for the given problem are moderate. For the example problem, the feedthrough approach needs about 83 % of the computational time of the damping approach.[4] When focusing on the linear part of the model, i.e., the structural model, the time savings are more pronounced. Relative to the

[3]The student work [S4] gives a more detailed analysis regarding the computational time in a linear system.
[4]The relative values are based on the average computational times of the models considered in this subsection.

total computational time with the damping approach, the structural model requires 10 % of the computational time with the damping approach and only 4 % with the feedthrough approach.

These savings do not only result from the faster solution of the linear structural model. Further time savings of the feedthrough approach are based on the fact that the artificial high-frequency modes are neglected dynamically. As a result, fewer oscillations occur and thus, fewer iterations are required for convergence.

Overall, the feedthrough approach shows better computational characteristics.

### Summary

An advantage of the damping approach is that it can be implemented relatively easy using modal damping. However, it has the disadvantage that the artificial modes are considered dynamically. Therefore, two parameters must be determined: $n_{\mathrm{dc}}$ and $\zeta$. For the choice of $n_{\mathrm{dc}}$, it has proven itself helpful to consider some artificial modes dynamically, i.e., to choose $n_{\mathrm{dc}} > n_q - n_{\mathrm{cm}}$. When selecting the damping ratio, the choice of $\zeta = 0.5$ proved to be sufficiently large.

The feedthrough approach has the advantage that the artificial modes are not considered dynamically and no parameter selection is necessary. However, because the structure of the system is not preserved, the implementation in commercial solvers is difficult.

The feedthrough approach has advantages over the damping approach in terms of computational time. The lower computational time is due to the fact that fewer degrees of freedom and less highly dynamic frequencies are taken into account in the time integration.

In summary, both approaches significantly reduce spurious oscillations compared to the untreated model, see Fig. 4.10. The advantages of the feedthrough approach prevail for the given problem.

Independent of the approach, the higher frequency dynamics of the system are not well approximated for the cases studied above, i.e., the cases with a cut-off frequency of $f_{\mathrm{c}} = 150$ kHz. The inferior approximation results from a too low cut-off frequency $f_{\mathrm{c}}$ of the model CB-150-CN. Figure 4.15 shows the results of models CB-150-CN and CB-450-CN using the feedthrough approach. The model with the higher cut-off frequency of $f_{\mathrm{c}} = 450$ kHz approximates the dynamic behavior of the force better. A further increase in the cut-off frequency does not further improve the accordance. Hence, in the following studies in this section, the model CB-450-CN combined with the feedthrough approach is used.

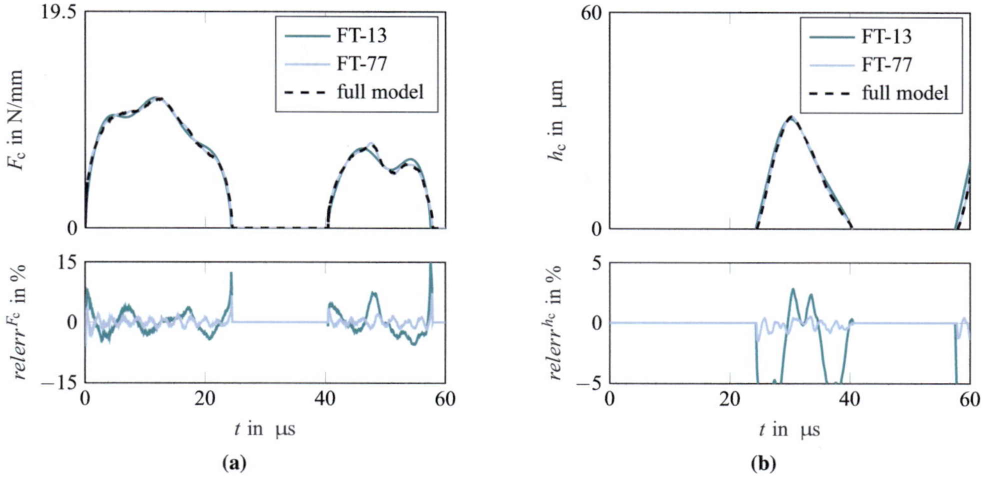


**Figure 4.15:** Results computed with the feedthrough approach for two reduced order models with different cut-off frequencies $f_c$ = 150 kHz and $f_c$ = 450 kHz. Since the cut-off frequency is higher, also more modes are considered dynamically, i.e., FT-13 and FT-77. The upper plots show the central force $F_c$ (a) and central gap height $h_c$ (b) over time $t$. The lower plots show the respective relative deviation of the reduced order model from the full model $relerr^{F_c}$ (a) and $relerr^{h_c}$ (b) over time $t$.

## 4.2.4 Time integration of the structural dynamics model

This work proposes the use of SDIRK methods for the fluid part as well as the structural part of the coupled problem. Section 4.1 investigated different time integration methods for the conventional EHD contact model and concluded that SDIRK methods are suitable to be used for the conventional EHD contact problem. As the extended EHD contact model captures further structural dynamic effects, it needs to be investigated whether the proposed time integration methods are suitable for the extended EHD contact model as well.

First, a brief overview of typical time integration methods in the field of structural dynamics is given, highlighting potential challenges. Thereafter, the application of SDIRK methods for problems in structural dynamics is addressed.

### Conventionally used time integration methods

The Newmark $\beta$-method with the parameters $\gamma = 0.5$ and $\beta = 0.25$ is a popular implicit time integration method for structural dynamics problems. It is second-order accurate and does not dissipate energy. However, for contact problems, spurious high-frequency oscillations can be observed, see [29].

To reduce these oscillations, numerical dissipation is introduced into the Newmark $\beta$-method by setting $\gamma > 0.5$. The introduction of dissipation commonly results in the loss of second-order accuracy. This loss of order motivated further research to identify and develop better suited time integration methods for structural dynamics, see [8, 49]. In this field, the Hilber-Hughes-Taylor method [71] and the generalized-$\alpha$ method [23] have evolved to become quite popular. These methods require the choice of parameters to adjust their properties for a given problem. An improper choice of parameters may result in divergence of the solution procedure.

Czekanski et al. [29] identified a set of optimal parameters for elastodynamic contact problems for the generalized-$\alpha$ method, with the goal to avoid high-frequency modes without dissipating energy. Based on these suggestions the parameters were adjusted to $\alpha = -0.05$, $\beta = 0.275625$ and $\gamma = 0.55$ for use with the HHT method [30]. This combination will be referred to as HHT-TF method throughout this work.

To illustrate the influence of time integration methods on the convergence behavior and solution quality, the example problem is solved with three different time integration methods. The methods compared are the Backward Euler method, the Newmark $\beta$-method with parameters $\gamma = 0.5$ and $\beta = 0.25$ and the HHT-TF method as described above.

The same problem parameters as in Subsection 4.2.1 are selected. The initial speed is set to be $v_0 = 1$ m/s and a constant time increment of $\Delta t = 60$ ns is chosen. The problem is solved using the commercial finite element software Abaqus.

Figure 4.16 shows the central force $F_c$ (a) and central gap height $h_c$ (b) resulting from the different time integration methods. The relative deviation $relerr$ of the compared method from the HHT-TF method is shown in the lower sub-plot. The relative deviation $relerr$ is computed using Eq. (A.4).

The NM method is unstable for the given problem, since oscillations occur that increase until the algorithm no longer converges for $t > 25$ μs. The Backward Euler method (BE1) introduces excessive damping, which is visible by the reduced rebound height. The HHT-TF method yields good results. These results show the influence of the time integration method and highlight the potential consequences of inappropriate parameter selections.

### SDIRK methods for problems in structural dynamics

Owren and Simonsen [103] study the use of SDIRK methods for problems in structural dynamics. Owren and Simonsen compare different SDIRK methods with the HHT method and conclude that the presented SDIRK methods are comparable or superior with respect to stability properties, relative period error and numerical cost. Additionally, they show better damping properties and allow for higher order accuracy, even in the case of a variable time step size. A further advantage is the possibility of error control through embedded methods, which is addressed in Section 2.4.

The study by Owren and Simonsen [103] investigates the methods in detail. In the scope of this work, only the convergence behavior is investigated, as it ensures the proper implementation and functionality of the methods. First, SDIRK methods, see Tab. 2.3, are applied to solve the equation of a simple harmonic oscillator. Thereafter, Alexander's second-order method is examined for the dry contact problem discussed above.

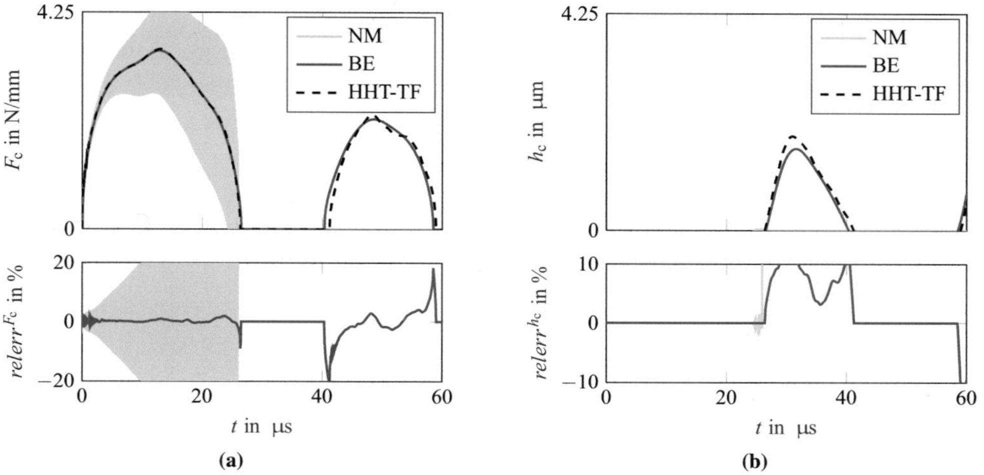


**Figure 4.16:** Results of the example problem with different time integration methods: Backward Euler method (BE1), the Newmark $\beta$-method (NM) and the Hilber-Hughes-Taylor method (HHT-TF). The upper plots show the central force $F_c$ (a) and central gap height $h_c$ (b) over time $t$. The lower plots show the respective relative deviation of the reduced order model from the reference solution $relerr^{F_c}$ (a) and $relerr^{h_c}$ (b) over time $t$.

*Convergence order study: simple harmonic oscillator*

The simple harmonic oscillator is a classical mechanics problem, for which analytical solutions exist. Since an analytical solution exists, it is particularly suitable for the investigation of temporal integration errors. The movement of a harmonic oscillator is described by

$$m\ddot{u}(t) + ku(t) = 0, \tag{4.3}$$

with the initial conditions $u(0) = 1$ m and $\dot{u}(0) = 0$ m/s, the analytical solution is

$$u_{\text{ref}}(t) = A\,\cos(\omega_0 t), \tag{4.4}$$

with the amplitude $A = 1$ m and the eigenfrequency $\omega_0 = \sqrt{k/m}$, where the stiffness $k$ and the mass $m$ are chosen to yield $\omega_0 = 2\pi$ 1/s.

Studying the convergence rate of a time integration method is helpful to see whether a method is functioning correctly for a given problem. Additionally, it is helpful in the context of this work since the adaptive time stepping algorithm is based on the convergence order of the method, see Subsection 2.4.2. For the convergence order study, Eq. (4.3) is solved with different time integration schemes and different time step sizes. Subsequently, the maximum absolute deviation $err_{\max}^{u}$ of the results from the analytical reference solution is computed using Eq. (A.3). The following study examines the backward Euler method (BE1), Alexander's second-order method (A2) and Hairer and Wanner's fourth order scheme (HW4), which is described in Tab. 2.3. The problem is solved for ten periods with fixed time step sizes $\Delta t = \{0.02\ \text{s},\ 0.01\ \text{s},\ 0.005\ \text{s},\ 0.0025\ \text{s}\}$.

Figure 4.17a shows the deviation $err_{\max}^{u}$ of each solution from the reference solution for the respective time step size $\Delta t$. The slope of the dashed line gives the theoretical convergence order of the respective method. The backward Euler method (BE1) does not reach its theoretical first order of convergence. This order degeneration results from its time step size dependent damping properties. However, A2 reaches its second order and HW4 reaches its fourth order. These results indicate the proper implementation of the time integration methods and support the results of [103] that SDIRK methods are suitable for structural dynamics problems.

*Convergence order study: dry contact problem*

The behavior of Alexander's second-order method for the fluid part of the problem is investigated in Section 4.1. In the following, the convergence behavior of the method for the structural dynamics part is examined. To this end, the dry contact problem is integrated in time using two different methods: the backward Euler method (BE1) and Alexander's second-order method (A2). The problem is solved with fixed time step sizes $\Delta t = \{240\ \text{ns},\ 120\ \text{ns},\ 60\ \text{ns},\ 24\ \text{ns}\}$. No analytical solution is given for this problem. Hence, a reference solution is computed numerically with a higher order scheme and a small time step size. The reference solution is computed using Hairer and Wanner's fourth order scheme (HW4) and a fixed time step size of $\Delta t = 6$ ns. Hairer and Wanner's fourth order scheme is described in Tab. 2.3. As for the simple harmonic oscillator, the maximum absolute deviation $err_{\max}^{h}$ from the reference solution is computed using Eq. (A.3). The problem parameters are given in Subsection 4.2.1. The initial impact velocity is set to $v_0 = 1$ m/s. The reduced order model CB-450-CN using the feedthrough approach is used as the structural dynamics model, see Subsections 4.2.2 and 4.2.3.

Figure 4.17b shows the resulting deviation $err_{\max}^{h}$ for each step size $\Delta t$. The slope of the dashed line indicates the theoretical convergence order of the respective method. As above, the damping of the Backward Euler method (BE1) leads to a degeneration of the convergence order. Alexander's second-order method (A2) reaches its theoretical second order. This result indicates a proper implementation and functioning of Alexander's second-order method for the given problem. These findings combined with the results of the studies of [103] suggest the suitability of SDIRK methods for the structural dynamics part of the EHD contact problem.

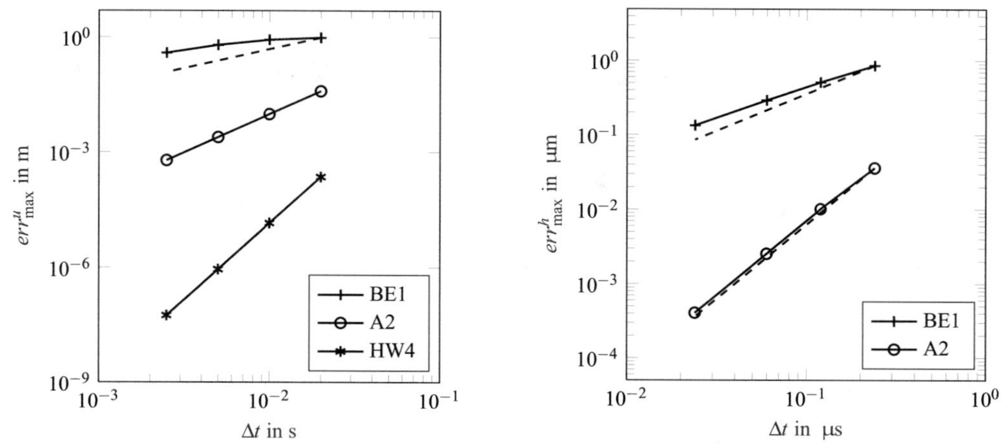


**(a)** Results for the simple harmonic oscillator.

**(b)** Results for the dry contact problem.

**Figure 4.17:** For each time step size and time integration method the deviation $err_{max}$ of the computed results from the reference solution is plotted over the step size $\Delta t$. The slope of the dashed line indicates the theoretical convergence order of the respective method.

## 4.3 The lubricated impact problem

When two bodies, which are separated by a lubricant, approach each other, high local fluid pressures arise in the contact area. Christensen [21] was the first to present theoretical and experimental studies of this problem in the field of lubricated contacts. His studies focused on the pressure build up as the surfaces approach each other. A constant approach velocity was assumed and the equation of motion was not solved. Dowson and Wang [38] and at around the same time Larsson and Höglund [85] were the first to generate a numerical solution for the impact and rebound of a sphere from a lubricated plate. These studies use the equation of motion of a point mass to compute the position of the sphere. In further studies, various other effects were investigated, see, e.g., [52, 82, 130]. However, to the author's knowledge, no extensions of the structural model have been proposed, which would enable the simulation of EHD contacts of complex bodies.

While the previous two sections investigated the fluid and the structural model in an isolated manner, this section investigates the completeextended EHD contact model. In this context, the solution procedure is applied to different example problems—all being variations of the axisymmetric lubricated impact problem.

First, a comparison against results from the literature is carried out to validate the extended model. In the scope of this comparison, aspects of the lubricated impact problem are highlighted.

The second subsection illustrates the influence of the solid's dynamic effects on the contact behavior. To this end, a valve-like geometry is examined. Through the variation of its material properties, differences to the conventional EHD contact model are highlighted.

The third subsection examines numerical aspects of the solution procedure. It investigates the influence of the coupling procedure on the spatial convergence order and discusses the computational cost based on the example problem.

While the example problems in the first three subsections deal with a single impact-rebound process, the final example problem studies the event of multiple successive impacts. To speed up the computation of multiple impact-rebound processes, model order reduction techniques are applied to the nonlinear coupled problem. In addition, techniques to reduce the effort involved in creating the reduced order model are proposed.

### 4.3.1 The impact of a sphere on a lubricated plate

The impact of a sphere on a lubricated plate is a typical example problem in the field of lubricated contacts. Experimental results by Safa and Gohar [114] as well as numerical results by Larsson and Höglund [85] are available for this problem. The availability of literature results make this problem setup suitable to be used for the validation of the extended EHD contact model.

In the following, the problem setup is outlined. Thereafter, the problem is solved with the extended EHD contact model and the impact-rebound process is discussed. Finally, the proper functioning of the extended EHD contact model is validated by comparison with literature results.

**Table 4.6:** Parameters of the axisymmetric lubricated impact problem based on the experimental setup of [114].

| parameter | $R_{\text{red}}$ | $v_0$ | $h_0$ | $E'$ | $m_0$ | $\eta_0$ | $\alpha_\eta$ |
|---|---|---|---|---|---|---|---|
| unit | mm | m/s | µm | GPa | kg | Pa·s | 1/GPa |
| value | 12.7 | 0.313 | 30 | 112.8 | 0.066 | 0.48 | 20 |

#### Problem setup

The same parameters as in the experimental study by Safa and Gohar [114] are used here. Figure 4.18 shows the problem setup. The steel sphere approaches the lubricated glass plate with the initial velocity $v_0$. It is accelerated by gravity with $g = 9.81\ \text{m/s}^2$, which, however, has a negligible influence on the contact force. The lubricant is modeled as compressible, see Eq. (2.10), and the pressure dependent viscosity variation is described by Roelands' model, see Eq. (2.9). Table 4.6 summarizes the parameters of the problem setup.

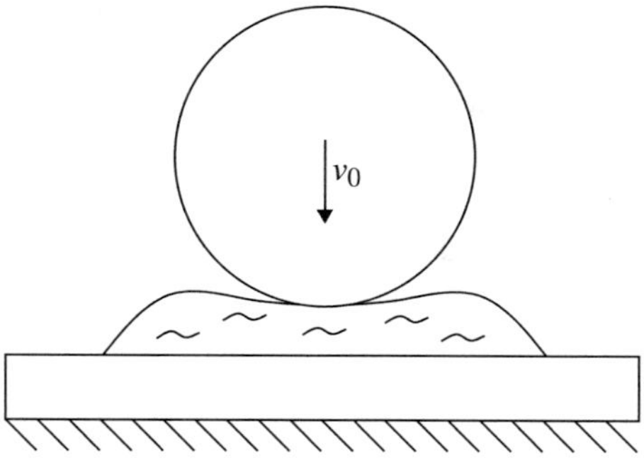


**Figure 4.18:** Schematic of the axisymmetric lubricated impact problem. The sphere moves with initial velocity $v_0 = 0.313$ m/s from the initial height $h_0 = 30$ µm towards the plate.

The sphere has a radius of $R = 12.7$ mm and its material properties are $E_2 = 210$ GPa, $\nu_2 = 0.3$, and $\rho_2 = 7692.1\ \mathrm{kg/m^3}$. The glass disk has a diameter of 150 mm and is 12.7 mm thick. The material properties of the plate are $E_1 = 70$ GPa, $\nu_1 = 0.25$, and $\rho_1 = 2500\ \mathrm{kg/m^3}$. The bodies are modeled in the commercial finite element software Abaqus using quadratic axisymmetric elements. The sphere is unconstrained and the plate is fixed at the bottom. Craig-Bampton method is used to construct a reduced order model with a cut-off frequency of $f_c = 150$ kHz and 301 nodes in the contact region (from 0 mm to 1.2 mm). The cut-off frequency of $f_c = 150$ kHz is sufficient in this case since the upper body is a sphere and the lower body a fixed plate. The computational domain is defined as $0 \leq X \leq 3.1$ and is divided into 502 intervals, using the grid ratio $i_{GR} = 0.6$, see Section 3.5. Alexander's second-order method is used for time integration with adaptive step size selection.[5]

### The impact-rebound process

Figure 4.19 shows the pressure (thick line) and the gap height (thin line) in the contact region at different time steps during the impact phase (upper row) and the rebound phase (lower row). As the sphere moves towards the plate, the gap height decreases and the pressure increases due to the squeeze term. The viscosity increases exponentially with increasing pressure. The viscosity increases so much that it enters the piezoviscous regime and the lubricant hardly flows. While the sphere continues to move towards the plate, the central gap height $h_c$ remains almost constant and the sphere deforms around the solidified lubricant. A dimple forms and the minimum gap height $h_{min}$ shifts to the periphery of the contact area where the lubricant is still flowing. The

[5]The adaptive time stepping uses the step size control (2.80) with absolute and relative tolerances $\epsilon_a^p = 1$ MPa, $\epsilon_a^u = 0.1$ nm, $\epsilon_a^q = 0.01$, and $\epsilon_r = 1 \cdot 10^{-4}$, for pressure, displacement and rigid body modes, respectively. Furthermore, the parameters $f_{min} = 0.25$, $f_{max} = 2$, and $f_{safety} = 0.6$ are used.

deviation between minimum gap height $h_{min}$ and central gap height $h_c$ is visible in Fig. 4.20b. The central gap height $h_c$ remains almost constant for most of the impact-rebound process.

When the sphere starts to rebound from the plate, the pressure initially decreases in the outer contact area and the high pressure area becomes smaller. Pressure peaks form in the outer contact area and move towards the center. During this phase, the maximum pressure $p_{max}$ is no longer in the center of the contact, but in the outer contact area. As the sphere leaves the plate, the outer pressure peaks merge at the contact's center, forming a second central pressure peak, see Fig. 4.20a.

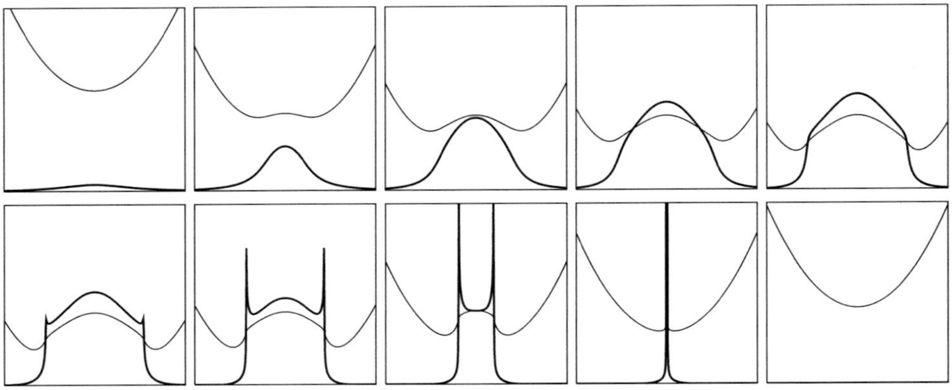

**Figure 4.19:** Temporal evolution of the pressure (thick line) and the gap height (thin line) spatially resolved in the contact region. The upper row shows the impact phase and the lower row shows the rebound phase.

The lubricated impact problem is strongly influenced by the transition of the lubricant into the piezoviscous regime, i.e., the point when the lubricant solidifies and stops flowing. The transition point is significantly influenced by the pressure-viscosity-coefficient $\alpha_\eta$ and by the approach velocity $v_0$. Hence, these two parameters—among others—strongly influence the characteristics of the lubricated impact problem. A variation of the pressure-viscosity-coefficient $\alpha_\eta$ influences the central gap height, the pressure distribution and the characteristics of the pressure peaks. A larger pressure-viscosity-coefficient results in a larger central gap height and larger pressure peaks. Similarly, an increase in the approach velocity $v_0$ leads to a faster build-up of pressure and thus to a faster solidification of the lubricant and consequently to a larger central gap height. The understanding of these relations is vital to gain an understanding of the lubricated impact problem. Summing up, the viscosity strongly influences the problem's characteristics. The works [52, 81, 125, 129, 130] give further insights into the lubricated impact problem.

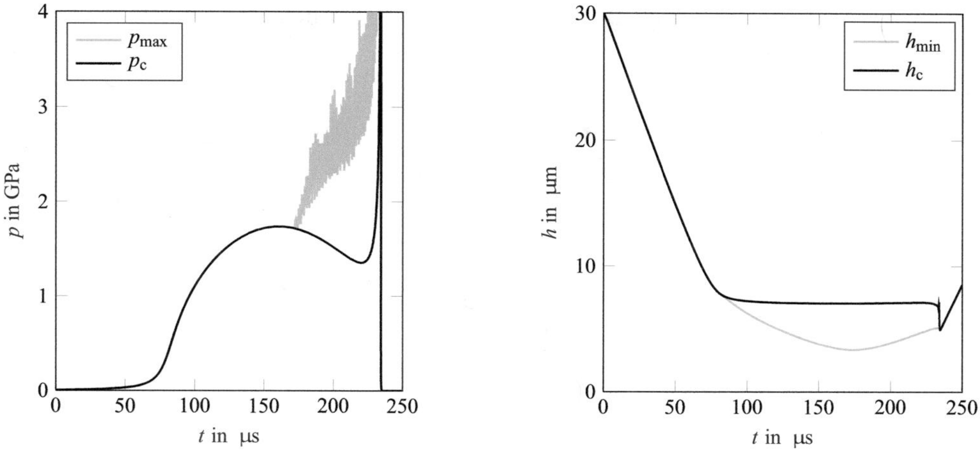


**(a)** Central pressure $p_c$ and maximum pressure $p_{max}$ over time $t$.

**(b)** Central gap height $h_c$ and minimum gap height $h_{min}$ over time $t$.

**Figure 4.20:** Gap height and pressure at discrete locations over time.

## Comparison with literature results

The above results are computed with the extended EHD contact model. In addition, experimental results from Safa and Gohar [114] and numerical results from Larsson and Höglund [85] exist. To validate the proper functioning of the proposed solution procedure and model, the current results are compared with literature results.

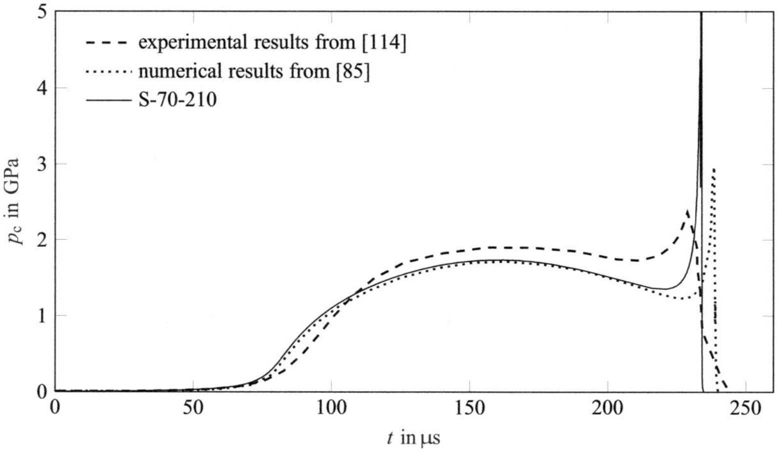


**Figure 4.21:** Central pressure $p_c$ over time $t$. The results of the developed extended EHD contact model (S-70-210, see Tab. 4.7) are compared with numerical results [85] and experimental results [114].

Figure 4.21 shows three different central pressure traces. The overall accordance is good, which validates the proper functioning of the extended EHD contact model. However, there is a slight difference between all three results.

The deviation from the experimental results from Safa and Gohar is caused by multiple factors. One part of the deviation results from the experimental and measurement setup while another part results from an insufficient physical model of the EHD contact.

The deviation from the numerical results from Larsson and Höglund is likely due to two reasons: the different numerical techniques and the different structural model, which are used in the presented procedure.

The higher pressure peak in the current results can be explained by the smaller time step size resulting from the adaptive time stepping scheme. Larsson and Höglund [85] use a fixed time step size of $\Delta t = 1$ µs and Backward Euler time integration. Here, an adaptive time stepping scheme is used. Figure 4.22 shows the step size $\Delta t_j$ over time $t$. The step size control reduces the step size during the rebound, thus, approximating the pressure peak in greater detail.

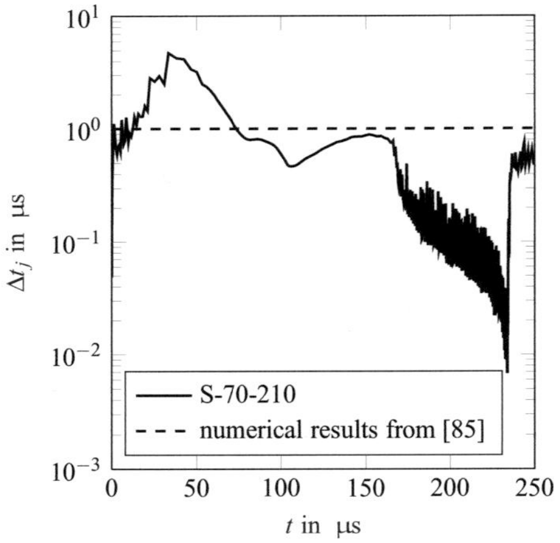


**Figure 4.22:** Time step size $\Delta t_j$ of the current model (S-70-210) and of [85]. During the early impact phase, a large step size $\Delta t_j$ is sufficient to approximate the pressure build up and motion of the sphere. During the rebound phase, pressure peaks move through the contact, see Fig. 4.19, and the step size is reduced significantly to capture these. After the rebound (at around 236 µm), the sphere flies freely and the step size increases rapidly.

The extended structural model used in the extended EHD contact model also leads to differences. Larsson and Höglund approximate the deformation with an elastic half-space approach and dynamic effects are solely considered by multiplying the sphere's mass with the acceleration acting on its center of gravity. This structural model will be referred to as the *conventional structural model*. It is summarized in Appendix B. It does not consider further structural inertia effects of the bodies.

Despite the neglection of dynamic effects, the results of the conventional structural model are in good accordance with the experimental results. This good accordance can be explained by the fact that the impacting body is a sphere with a large Young's modulus. For such compact bodies, a simple structural model is usually sufficient, see [46, 121]. However, the model is no longer sufficient if complex geometries need to be considered.

### 4.3.2 The impact of a valve-like geometry on a lubricated plate

Problems with compact bodies can be studied adequately using the conventional EHD contact model. For more complex non-compact geometries, however, the conventional model does not yield useful results, since it neglects the structural dynamic behavior of the bodies. This characteristic limits the applicability of the conventional EHD contact model.

The following study investigates the axisymmetric lubricated impact problem of a valve-like geometry. To construct different setups, the material properties of the lower and upper body are varied. However, the reduced material properties, i.e., $E'$, $R_{\text{red}}$, and $m_0$, are the same in all setups. The behavior of the different setups is investigated using the extended EHD contact model.

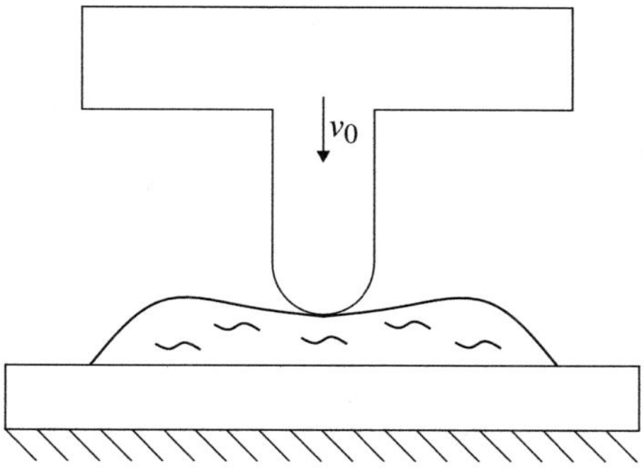


**Figure 4.23:** Exemplary drawing of the impact of the axisymmetric T-geometry on a lubricated plate. The T-geometry moves with initial velocity $v_0$ = 0.313 m/s from the initial height $h_0$ = 30 µm towards the plate.

#### Problem setup

Figure 4.23 shows the problem setup. The geometry of the upper body is motivated by the shape of a valve needle as it can be found in injectors. The T-geometry is designed such that it has the same volume as the sphere in the example problem above, see Subsection 4.3.1. It is shaped spherically in the contact area, such that the problem setup has the same reduced radius $R_{\text{red}}$, see Eq. (2.18), as in the example problem above. Also, the plate has the same dimensions as above. The density of both bodies is chosen as in the example problem above, such that the upper body has the same mass $m_0$. Finally, also the elastic material properties are chosen such that they give the same resulting Young's modulus $E'$, see Eq. (2.17), as in the example problem above. However, the combination of the elastic material properties of the lower and upper body is different.

Three additional setups are created in which the combination of the elastic material properties of the lower and upper body is varied.[6] The elastic material properties of the setups are given in Tab. 4.7. The fluid parameters are the same as in the example problem above. Hence, all parameters influencing the conventional EHD contact model are the same as in Subsection 4.3.1. Only the shape and elastic material properties of the bodies differ. Note that, if the four setups were computed with the conventional structural model, all setups would yield the same results as for the sphere on plate example problem.

**Table 4.7:** Elastic properties of the different setups. $E_1$ and $\nu_1$ refer to the lower body (plate) whereas $E_2$ and $\nu_2$ refer to the upper body. In the first three setups the upper body is the T-geometry, therefore, the setup ID starts with T. In the fourth setup the upper body is a sphere, hence, the setup ID starts with S. The fourth setup is the same setup as in Subsection 4.3.1. Note that, the reduced material properties are the same in all setups, i.e., $E' = 112.8$ GPa, $R_{\text{red}} = 12.7$ mm, and $m_0 = 0.066$ kg.

| setup ID | $E_1$ | $\nu_1$ | $E_2$ | $\nu_2$ |
|---|---|---|---|---|
| T-70-210 | 70 GPa | 0.25 | 210 GPa | 0.3 |
| T-103-103 | 102.7 GPa | 0.3 | 102.7 GPa | 0.3 |
| T-210-70 | 210 GPa | 0.3 | 70 GPa | 0.25 |
| S-70-210 | 70 GPa | 0.25 | 210 GPa | 0.3 |

### Results

Figure 4.24 shows the central pressure $p_c$ (a) and central gap height $h_c$ (b) over time $t$ for the four different setups, see Tab. 4.7. The results are computed using the extended EHD contact model. All setups have the same reduced parameters, see Tab. 4.7. The figures show differences between the results for the sphere (S-70-210) and the three T-geometry setups. The different behavior of the T-geometry in the contact region results from its dynamic characteristics outside the contact region. The deformations of the horizontal bar of the T-geometry influence the contact force over time. With increasing flexibility of the body, this effect becomes more pronounced, i.e., with increasing flexibility of the T-geometry, the contact time increases. Additionally, small oscillations can be observed. The oscillations result from the vibrations excited by the rapid force increase during the deceleration. After the rebound, the T-geometry continues to vibrate, which can be observed for setup T-70-210 in Fig. 4.24b.

Since the conventional structural model neither takes into account non-local deformations nor oscillation behavior of the bodies, the conventional structural model gives the same result for all

[6]The setups are created in the finite element software Abaqus using the same approach as in Subsection 4.3.1.

considered setups. The extended structural model, on the other hand, takes these effects into account. Therefore, the extended structural model can also be used for applications with more complex geometries, such as valves.

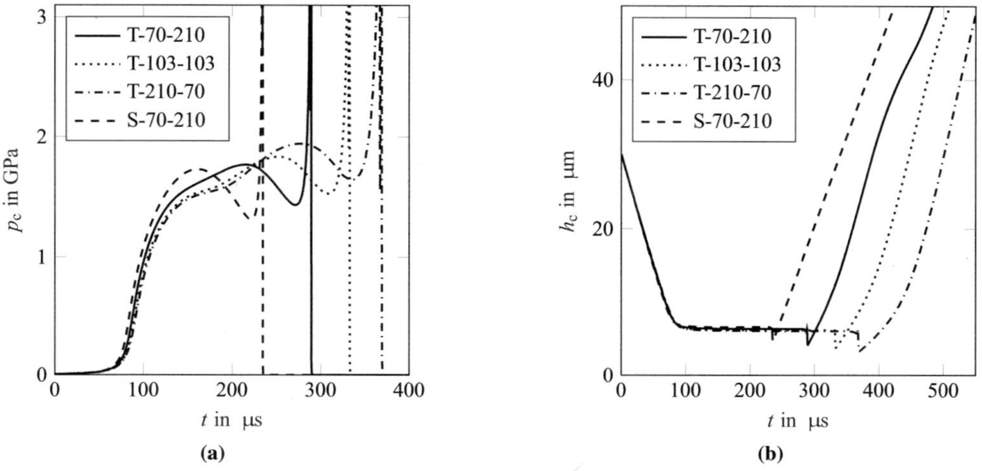


**Figure 4.24:** Central pressure $p_c$ (a) and central gap height $h_c$ (b) over time $t$. Table 4.7 lists the elastic material properties of the four setups.

## 4.3.3 Remarks on the solution procedure

The extended EHD contact model considers further effects compared to the conventional EHD contact model. However, an extension of the modeling is usually accompanied by an increase in the computational cost. Additionally, here, a fluid-structure coupling is introduced to connect the domains. The solution procedure, developed throughout this work, aims to keep the additional computational effort as low as possible. This procedure is explained in Chapter 3 and examined throughout this chapter. In the following, two aspects of the solution procedure are highlighted: the coupling procedure and the additional computational effort needed for the given problem.

### Coupling procedure

To obtain the extended structural model, the elastic bodies are first discretized using finite elements. Reduced order models are created from the finite element models. The reduced order models approximate the body's behavior at the nodes in the contact region. These nodes form the structure grid. The fluid equations are solved on a different grid. These grids are not necessarily equal, thus, three different grids may exist. The displacements of the bodies are given on the structure grids and the pressures are given on the fluid grid. To solve the coupled problem, the physical quantities of the different domains must be connected. Section 3.5 explains the coupling procedure.

**Table 4.8:** Studied grid combinations. $n_{S[1]}$ and $n_{S[2]}$ give the number of structure nodes of the lower and upper body, respectively. $n_F$ gives the resulting number of fluid nodes, depending on the prescribed grid ratio $i_{GR}$.

| case ID | $n_{S[1]}$ | $n_{S[2]}$ | $n_F\,(i_{GR}=1)$ | $n_F\,(i_{GR}=0.6)$ |
|---|---|---|---|---|
| 1 | 131 | 151 | 153 | 253 |
| 2 | 281 | 301 | 303 | 503 |
| 3 | 541 | 601 | 603 | 1005 |
| 4 | 801 | 801 | - | 1339 |

The coupling procedure is constructed in such a way that it supports elements with quadratic shape functions. The fluid equations are discretized with second-order finite differences. This study examines whether the spatial order of the discretization schemes is maintained after coupling.

To study the spatial convergence order of the coupled problem, finite element models with different discretizations in the contact region are created. Based on the structure grids, the respective fluid grid is constructed according to Eq. (3.26) using the grid ratio $i_{GR}$. Table 4.8 lists the number of nodes in the contact region for the studied structure grids and the respective fluid grid with $i_{GR}=1$ and $i_{GR}=0.6$. In all cases, the lubricated impact problem of the sphere is solved using a constant time step size of $\Delta t = 50$ ns and Alexander's second-order method for time integration. The solution of case 4 is used as the reference.

The deviation $err_{max}$ of the solution from the reference solution is computed for the impact phase, i.e., $t < 165$ µs. Only the impact phase is investigated, since, during the rebound phase, independent of the coupling, the second order is not achieved due to the grid dependent pressure peaks. The deviation $err_{max}$ is computed using Eq. (A.3).

Figure 4.25 shows the decreasing deviation $err_{max}$ with increasing number of fluid nodes $n_F$. The dashed line indicates an error decay of second order. The computed deviations are in close agreement with the theoretical error decay. It can be concluded that the error introduced by the coupling procedure is small. Furthermore, it can be concluded that the coupling procedure can maintain the order of the spatial discretization scheme.

### Computational efficiency

Chapter 3 presents measures to keep the additional computational effort for the extended model low. It is difficult to make a general statement regarding the required computational cost, however, since it depends, among other things, on the discretization, number of modes, number of iterations, the programming language and efficiency of the implementation. However, in the following the computational cost of the extended EHD contact model for the sphere on lubricated plate example problem is examined. In this example problem, the system consists of 503 fluid nodes, 301 upper body nodes, 3 upper body modes, 281 lower body nodes, and 83 lower body modes. For comparison, the same problem with the same number of fluid nodes is computed using the conventional structural model. The extended structural model needs about 8 % more computational effort compared to the conventional structural model.

This result shows, that the proposed solution procedure, as presented in Chapter 3 and studied throughout this chapter, minimizes the additional effort introduced by the extended structural model. Independent of the modeling approach, for some applications, a further reduction of the whole coupled problem can be useful. The application of a further model order reduction to the whole coupled problem is studied in the following.

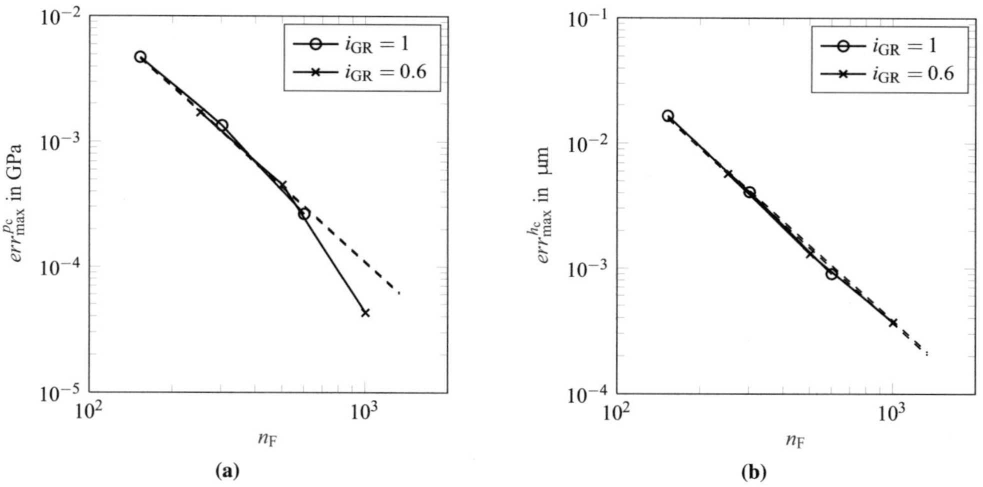


**Figure 4.25:** Investigation of the spatial convergence order after the coupling for the central pressure $p_c$ (a) and for the central gap height $h_c$ (b).

### 4.3.4 Efficient computation of multiple impacts

Lubricated contact simulations can be used to estimate the wear of components. To this end, similar processes are successively simulated multiple times. These simulations are often time-consuming. Model order reduction (MOR) techniques can be applied in order to reduce the computational effort needed to compute similar processes.

Maier [93] applied MOR techniques to the EHD contact problem for entraining motion contacts. The whole system of equations was reduced, i.e., both the nonlinear fluid model and the linear structural model. In the following, MOR techniques are applied to the extended EHD contact model in order to reduce the computational effort of the lubricated impact problem. Section 3.7 discusses the reduction procedure.

In this work, the fundamentals of the reduction procedure as given in [93] are used. However, some extensions are made in order to improve efficiency:

- Adaptive ROM construction: To minimize the offline phase of the MOR procedure, the ROMs are constructed during the solution process from already computed solutions.
- Local ROMs: Amsallem et al. [2] presented a MOR approach using local reduced bases. In this work, a similar procedure is introduced to the lubricated impact problem using the extended EHD contact model. To this end, different local ROMs are computed for different parts of the transient solution.

The reduction procedure is applied to speed up the computation of multiple successive impacts. The computations are carried out with the extended EHD contact model. The effectiveness of the reduction procedure is examined on the basis of computational time and accuracy.

### Adaptive reduced order model construction

To construct a ROM, computational effort is required. The construction process for nonlinear models usually consists of two steps. First, training solutions are computed. It is essential to choose the training solutions such that they contain enough information to cover the desired parameter space. Frequently, for example, parameter variations are carried out in the desired parameter room. Second, the ROM is constructed from the training solutions as outlined in Section 3.7.

The time needed to compute the training solutions and to construct the ROM is referred to as offline time. The time it takes to compute the problem with the ROM is called online time. A prior investment of offline time is necessary to benefit later from the shorter online time. Depending on the effort to compute the training solutions, this time investment can become large. Therefore, before creating a ROM, it must be ensured that the application of MOR techniques is helpful for the given problem.

Here, the problem of multiple lubricated impacts is studied. This problem motivates the idea to use solutions that need to be computed in any case as training solutions for the ROM. Hence, a procedure is proposed which automatically constructs a ROM from the solutions computed so far and, if possible, uses the ROM in the further solution process.

To ensure a certain quality of the ROM solution, an error estimate is computed in each time step. Here, the root mean square of the residuum of the nonlinear function, see Eq. (3.48), is used as an error estimate, i.e., $err_{\mathrm{ROM}} = \mathrm{rms}\left(\boldsymbol{f}_p\right)$. If the error estimate $err_{\mathrm{ROM}}$ exceeds a predefined tolerance $tol_{\mathrm{ROM}}$, a full system solution is computed for that time step. This procedure ensures a certain solution quality while introducing little additional computational time. The flowchart in Fig. 4.27 outlines this procedure.

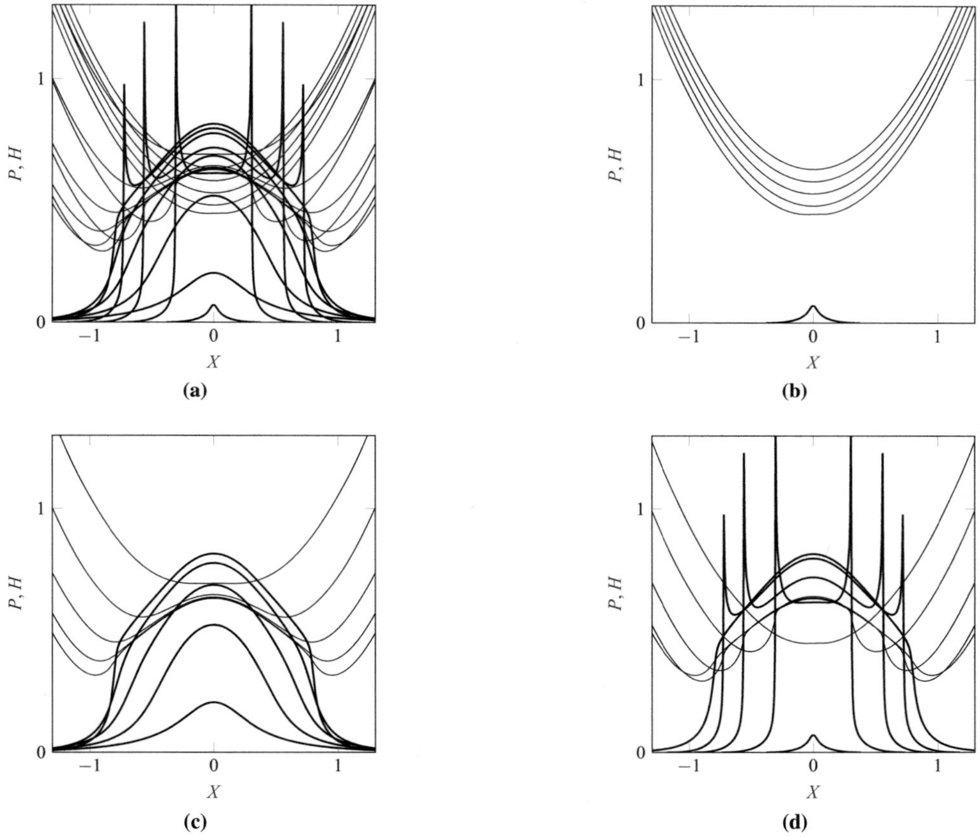


**Figure 4.26:** Temporal evolution of the pressure (thick line) and the gap height (thin line) in the contact region, during the phases of the impact-rebound process. The similarity of pressure and gap height solution snapshots throughout the entire process (a) is rather low. By dividing the process into a flight phase (b), impact phase (c), and rebound phase (d), a greater similarity is visible.

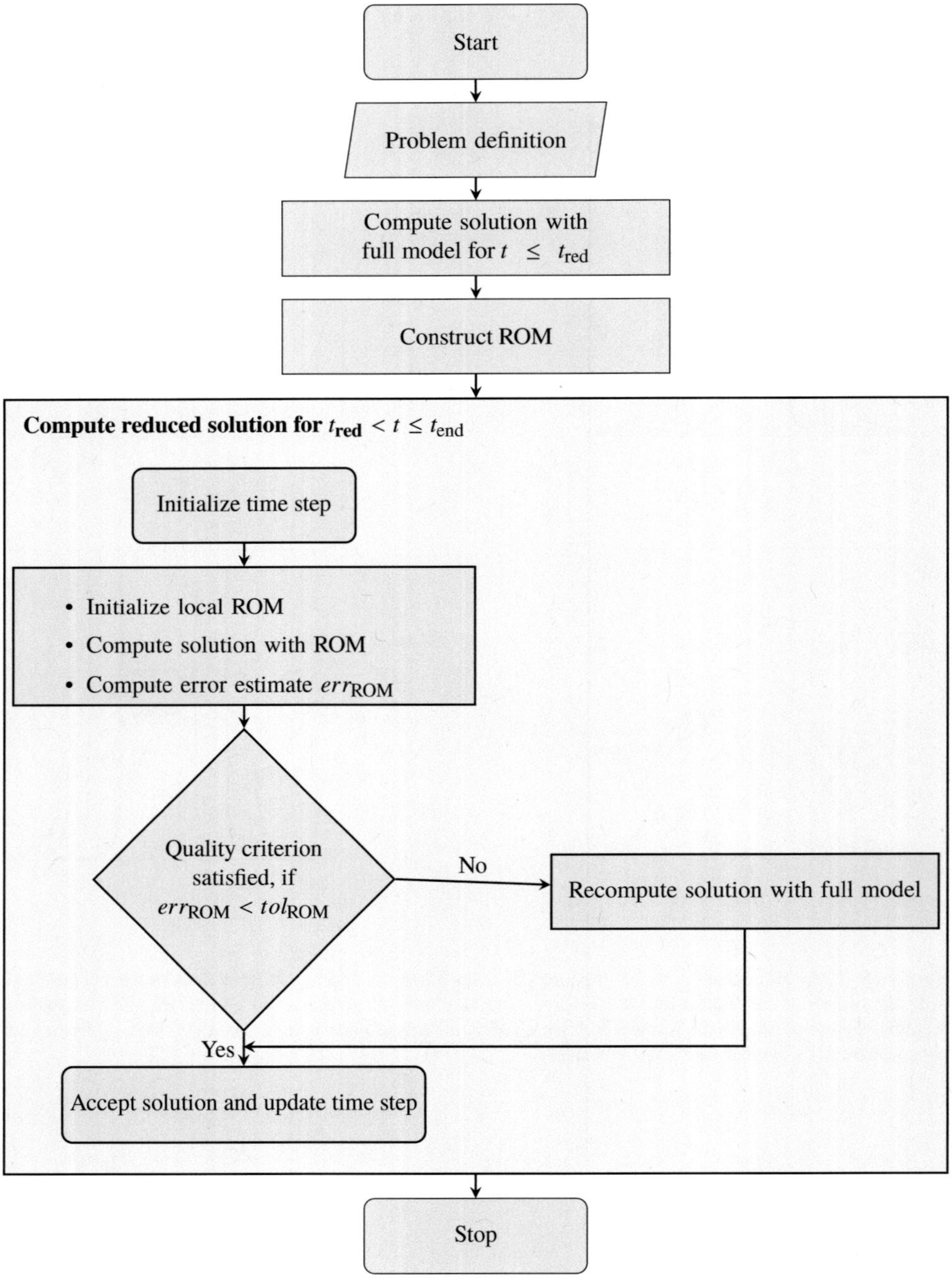


**Figure 4.27:** Outline of adaptive-local ROM computing procedure.

### Local reduced order models

Commonly, MOR procedures use one global ROM for the entire solution process. The size and quality of a ROM depend on the similarity of the solutions, which are used to construct the ROM. Increasing the similarity of the solutions may yield a reduction of the dimension of the ROM while maintaining the same accuracy. The similarity can be increased by identifying resembling phases in the solution process and creating specific ROMs for these phases.

The temporal evolution of the lubricated impact problem can be split into three phases: the flight phase, the impact phase, and the rebound phase. Figure 4.26 shows the solution traces of the pressure and the gap height at different times during one impact-rebound process (Fig. 4.26a) and during the three different phases, see Figs. 4.26b through 4.26d. The similarity of the solutions is higher during the individual phases as compared to during the complete impact-rebound process.

The similarity of the solutions can also be examined by observing the decay of the loss value. The loss value is used as an indicator of the information content of a ROM. It is computed according to Eq. (3.60). A steep decay of the loss value indicates a similarity of the solutions. In this case, few degrees of freedom of the ROM suffice to approximate the solution.

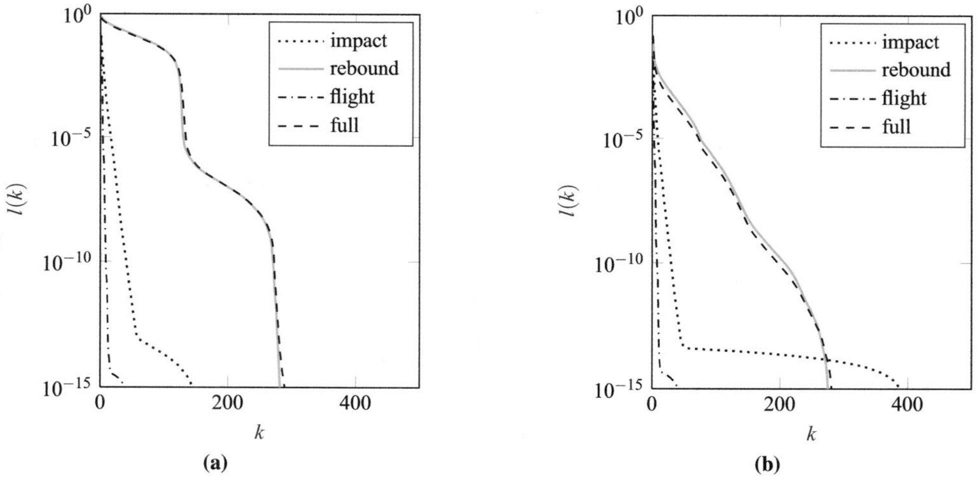


**Figure 4.28:** Loss value $l(k)$ over system size $k$ for the pressure (a) and for the gap height (b). The different decay rate of the loss value in the different phases indicates the difference in the similarity of the solutions in the different phases. A steeper decay typically results in a smaller system size.

Figure 4.28 shows the loss value $l(k)$ over the system size $k$ for the complete impact-rebound process and the different phases. For the impact phase and the flight phase, the loss value decays rapidly. The rapid decay indicates a high similarity of the solutions. Hence, a ROM with only a few degrees of freedom can cover this phase of the problem.

The loss value of the rebound phase and that of the complete impact-rebound process decays slowly. Hence, more degrees of freedom are needed to capture relevant information. The lower similarity of solutions in this phase can be explained by the pressure spikes moving through the contact region. Overall, the observations show that a small ROM would suffice to describe the impact and flight phase and a large ROM is needed to describe the impact-rebound phase.

Generating a single ROM for the complete process would require a large system size. However, most of the information content of such a ROM would only be required for the rebound phase. The rebound phase accounts for less than half of the process. For this reason, local ROMs are introduced here. The application of local ROMs enables the use of smaller ROMs for different parts of the solution process.

**Example problem**

The problem of a sphere impacting a lubricated plate is investigated. The same problem setup as in Subsection 4.3.1 is used. To accelerate the event of multiple impacts, the vertical acceleration acting on the sphere is increased to $a_{\mathrm{v}} = 757.58\ \mathrm{m/s^2}$, which is equivalent to a load of 50 N acting vertically on the sphere. Due to the short contact time and the large contact force, this acceleration has a negligible influence on the contact behavior. However, since it acts on the body during the flight phase, it accelerates the event of multiple impacts. The simulated time interval is chosen to be $0\ \mu\mathrm{s} \leq t \leq 4000\ \mu\mathrm{s}$. This interval is large enough to observe multiple impacts. The ROMs are computed after $t_{\mathrm{red}} = 250\ \mu\mathrm{s}$, which is approximately the duration of the first complete impact-rebound process. Two local ROMs are constructed: one for the flight phase, i.e., $\max(P) \leq 0.2$, and one for the impact and rebound phase, i.e., $\max(P) > 0.2$.[7] The maximum tolerable loss value is chosen to be $l_{\max} = 1 \cdot 10^{-7}$. This value yields a system size for the local ROM approach of $\widetilde{n}_z = 34$ in the flight phase and of $\widetilde{n}_z = 319$ in the impact and rebound phase. The global ROM has a system size of $\widetilde{n}_z = 316$ and the unreduced reference model has a system size of $n_z = 1004$. This clearly shows the lower number of degrees of freedom resulting from the reduction procedure. The tolerance criterion is set to $tol_{\mathrm{ROM}} = 1$.

Figure 4.29 compares the central pressure $p_{\mathrm{c}}$ and central gap height $h_{\mathrm{c}}$ obtained with the adaptive-local ROM to the unreduced reference solution.

[7] Using three local ROMs was also investigated. However, at the transition point from the impact to rebound phase a jump in the solution occurred due to the different information content of the local ROMs. This jump introduced oscillations in the dynamic problem yielding more rejected solutions and a longer overall computational time compared to using only two local ROMs.

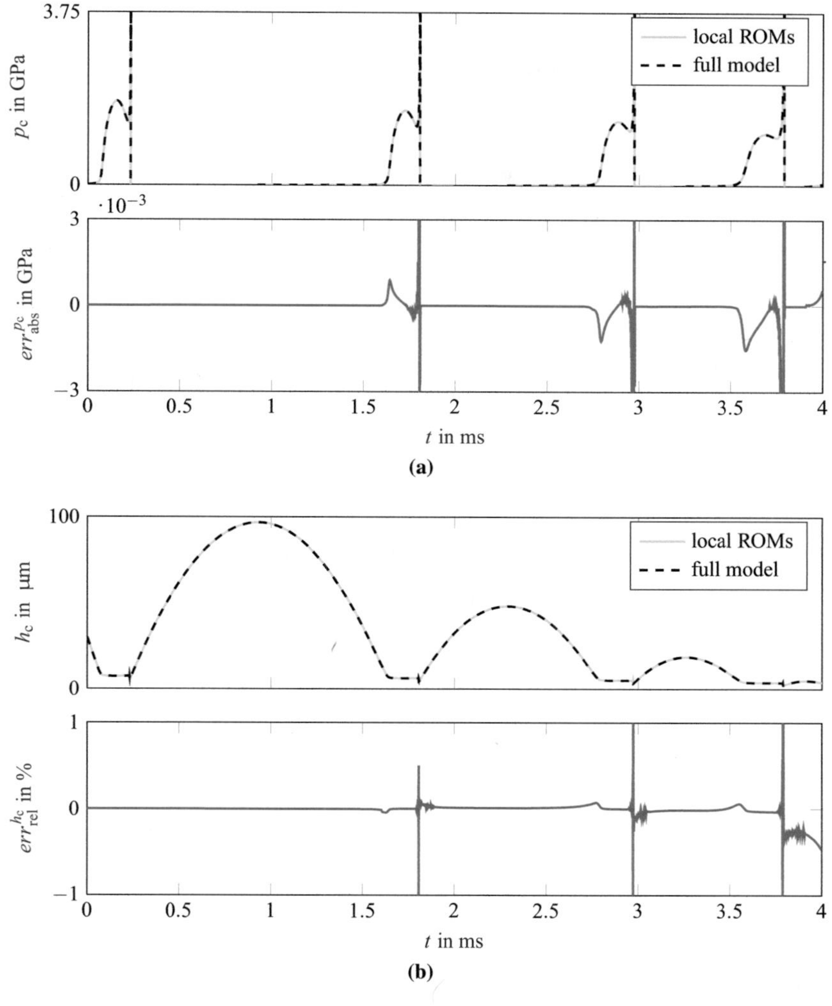


**Figure 4.29:** The upper plots show the central pressure $p_c$ (a) and central gap height $h_c$ (b) over the time $t$. The lower plots show the respective absolute deviation of the reduced order model from the full model $err_{abs}^{p_c}$ (a) and $err_{abs}^{h_c}$ (b) over the time $t$.

For the central pressure $p_c$, the absolute deviation is computed as

$$err_{abs}^{p}(t) = p_c^{ref}(t) - p_c^{ROM}(t),$$

and for the central gap height $h_c$, the relative deviation is computed as

$$err_{rel}^{p}(t) = \frac{h_c^{ref}(t) - h_c^{ROM}(t)}{h_c^{ref}(t)}.$$

**Table 4.9:** This table lists the relative computational time, system size $\widetilde{n}_z$, and number of evaluation points $\widetilde{N}_p$ of the investigated models.

| model | rel. comp. time | $\widetilde{n}_z$ impact/rebound | $\widetilde{n}_z$ flight | $\widehat{N}_p$ |
|---|---|---|---|---|
| local ROM | 0.51 | 319 | 34 | 290 |
| global ROM | 0.59 | 316 | 316 | 290 |
| full | 1 | 1004 | 1004 | - |

During the pressure build-up and around the pressure spike, a small deviation can be observed. Overall, the deviation is minimal. Table 4.9 summarizes the system size and relative computational time of the different models. The adaptive-global approach reduces the computational time to 59 % of the full model's computational time. Combining the adaptive approach with local ROMs, yields a reduction to 51 % of the full model's computational time.

The speedup resulting from MOR techniques depends strongly on the studied problem. Here, the studied problem is nonlinear, which results in many evaluation points and a large system size. The local ROM approach can reduce the system size in parts of the solution process. The reduced system size yields a further reduction of the computational time. Summing up, the adaptive ROM approach allows the application of MOR techniques to problems with recurring events, reducing total computational time by about half while minimizing additional prior computational cost.

# 5 Summary and outlook

The objective of this work is to develop an efficient numerical solution procedure for the lubricated impact problem which captures structural dynamic effects of complex geometries. For this purpose, the conventional EHD contact model is extended with respect to the structural model. Numerical techniques, such as model order reduction and adaptive time integration, are used to obtain a low computational time.

To consider the geometry of the bodies, the contact bodies are discretized using finite elements. This modeling approach makes it possible to represent more complex shapes and takes structural dynamic behavior into account. In order to decrease the number of resulting degrees of freedom, the finite element models are reduced using model order reduction techniques. For the reduction of the linear finite element models, component mode synthesis is used. Piezoviscous and compressible fluid behavior is considered and fluid flow is approximated utilizing the nonlinear Reynolds Equation. An efficient procedure for the coupling of the two bodies with the fluid is presented, which yields stable results and allows to maintain the convergence order of the spatial discretization schemes. Due to the procedures used, the resulting equation system has the same size as the equation system of conventional EHD contact models. Time integration of the system is carried out using singly diagonal implicit Runge-Kutta (SDIRK) methods. These methods are particularly suitable for stiff problems and allow for efficient adaptive step size control based on embedded schemes. The fully coupled system is solved in a monolithic approach using the Newton-Raphson method. Finally, the Proper Orthogonal Decomposition method in combination with a system approximation is used to reduce the size of the entire nonlinear system. The proposed solution procedure is applied to different problems, which allow for a separate investigation of different parts of the solution procedure.

The shock loading of a lubricated contact is the first example problem. This problem is used to examine the proposed adaptive time integration procedure. In EHD contact problems the Backward Euler method or the second-order finite difference method is typically used for time integration. This work proposes the use of SDIRK methods, specifically Alexander's second-order method. The proposed method is compared with the conventionally applied methods using the above example problem. Furthermore, the time step size control is examined. The rapid load change in this problem is a challenge for the time step size control algorithm and causes rejected time steps. An extension of the time step size control is proposed, to reduce the number of rejected time steps. This extension takes into account the integration error of the parameter function in the step size selection. It requires almost no additional computational effort while reducing the number of rejected steps and thus the computational time.

The second example problem is the impact of an elastic body on a rigid plate. The body's shape is motivated by that of a valve-needle in a magnet injector. The body's geometry influences its behavior in the contact region. Using this example problem, the extended structural model is examined. The extended structural model is based on reduced order models that are constructed using component mode synthesis. Component mode synthesis methods can be divided into two main groups: fixed-interface and free-interface methods. The comparison of the two methods shows that the fixed-interface method is better suited for the examined application. However, without specific adaptations, spurious oscillations are observed independently of the method. Two approaches to prevent these oscillations are investigated: the feedthrough approach and an approach based on modal damping. The feedthrough approach yields better results. However, it is more difficult to implement. If the implementation is not feasible, the spurious oscillations can also be reduced using modal damping. Finally, the proposed time integration procedure is briefly examined for structural dynamics and shows good results.

The third example problem concerns with the lubricated impact problem. Unlike the first and second example problem, several sub-problems are studied for this one. For the study of the lubricated impact, the entire solution procedure is applied. The first sub-problem concerns with the impact of a sphere on a lubricated plate. This problem is used to validate the solution procedure by comparing results from the proposed procedure with experimental and numerical results from the literature. The results show good agreement. The second sub-problem concerns with the impact of a valve-like body on a lubricated plate. In this study, the elastic material properties of the contact bodies are varied. The material properties are varied in such a way that a conventional EHD contact model would yield the same solution for all cases. The extended EHD contact model, however, captures clear differences in the results of the different material combinations. These results show the necessity of the extended structural model for more complex geometries.

An extension of the modeling is usually accompanied by an increase of the computational cost. For this reason, computational aspects are investigated. Generalized statements on the computational efficiency of a procedure are difficult to make, since they strongly depend on the problem type, the number of degrees of freedom and other aspects. However, for the given problem, the extended EHD contact model requires only 8 % additional computational time compared to the conventional EHD contact model.

The third and final sub-problem concerns with multiple successive impacts. Such computations are used, for example, to estimate wear, which requires the repeated computation of a similar problem. To save computational time, here, the use of model order reduction techniques is suitable. Typically, the creation of reduced order models requires training computations in advance. These training computations increase the effort of reduced order model creation and thus increase the threshold for the efficient usage of reduced order models. In order to keep the effort for creating a reduced order model low, an adaptive reduced order model creation procedure is proposed here. The first part of the solution is computed with the unreduced model and a reduced order model is then built based on these results. This approach has the advantage that little additional effort is required to create the reduced model since the computation of the first part of the solution is necessary independently of the reduction. Furthermore, the use of local reduced order models is proposed, i.e., using different local reduced order models for different phases of the solution process. This approach allows for the use of smaller reduced order models for phases of the solution process.

The presented solution procedure takes further physical effects into account and requires only a small additional numerical effort. The use of model order reduction techniques allows a further reduction of computational time and the creation of reduced order models that can be used in system simulations. The flexible and accurate structure modeling coupled with the fast numerical solution approach enables the usage of the presented solution procedure in a wide range of applications.

# Outlook

In the lubricated impact problem high pressures occur. These high pressures can cause plastic deformation. In order to investigate these effects more closely, the material model could be extended to take plastic material behavior into account. Furthermore, plastics are increasingly used in lubricated contacts. To accurately model these, more complex material models are required.

Further efforts could also be made to consider thermal effects. The temperature has a significant influence on the viscosity, which in turn greatly influences the lubricated contact behavior. Both the fluid model and the structural model should be extended to include thermal effects. To enhance the structural model to include thermal effects, model order reduction techniques, such as those used in structural dynamics, could prove to be helpful.

As the gap height and impact velocity increase the consideration of inertial effects in the fluid can become relevant. With this extension, the equation system would become more complex. Hence, the suitability of other discretization methods should be investigated. As with all model extensions, the additional computational effort should be weighed up against the increased model quality.

In this work, only centric impacts are investigated. Non-centric impacts may influence wear and noise related problems. In further works, the model could be extended to allow the computation of non-centric impacts.

The solution procedure presented in this work is implemented in Matlab and represents a proof of concept. There is still the potential to accelerate the solution procedure by applying and extending existing approaches. These approaches should be investigated in the context of industrialization through implementation in other programming languages, such as C++. In the current implementation, the complete Jacobian matrix is built up in every iteration. The build-up of the Jacobian matrix takes much computational time. There are approaches to reduce this effort, for example by using the same Jacobian matrix for multiple iterations. Furthermore, there exist approaches to keep parts of the Jacobian matrix constant in combination with SDIRK methods. Within the framework of the implementation of this solution procedure, the applicability of such procedures should be examined.

# A Computation of absolute and relative deviations

The deviation between different quantities is computed in various ways throughout this work. The relevant formulas are listed below.

The root mean square (rms) of the vector $\boldsymbol{x} \in \mathbb{R}^n$ corresponds to a typical value of an element in $\boldsymbol{x}$

$$\mathrm{rms}\,(\boldsymbol{x}) = \sqrt{\frac{1}{n}\left(x_1^2 + x_2^2 + \cdots + x_n^2\right)}. \tag{A.1}$$

The deviation between a variable vector $\boldsymbol{x}_j$ at time instant $t_j$ and its reference $\boldsymbol{x}_j^{\mathrm{ref}}$ is computed using the following formulas,

$$relerr^{x}_{\mathrm{mean}} = \mathrm{rms}\left(\frac{\left\|\boldsymbol{x}_j^{\mathrm{ref}} - \boldsymbol{x}_j\right\|}{\left\|\boldsymbol{x}_j^{\mathrm{ref}}\right\|}\right) \cdot 100\,\%,\ \text{and} \tag{A.2}$$

$$err^{x}_{\max} = \max_{j:t_j \in [t_0,\, t_{\mathrm{end}}]} \sqrt{\frac{1}{n}} \left\|\boldsymbol{x}_j^{\mathrm{ref}} - \boldsymbol{x}_j\right\|. \tag{A.3}$$

The deviation between a scalar $x_j$ at time instant $t_j$ and its reference $x_j^{\mathrm{ref}}$ is computed as

$$[relerr^{x}]_j = \frac{x_j^{\mathrm{ref}} - x_j}{\max\limits_{j:t_j \in [t_0,\, t_{\mathrm{end}}]} \left|x_j^{\mathrm{ref}}\right|} \cdot 100\,\%,\ \text{and} \tag{A.4}$$

$$relerr^{x}_{\max} = \max_{j:t_j \in [t_0,\, t_{\mathrm{end}}]} \left(\frac{x_j^{\mathrm{ref}} - x_j}{\max\limits_{j:t_j \in [t_0,\, t_{\mathrm{end}}]} \left|x_j^{\mathrm{ref}}\right|} \cdot 100\,\%\right). \tag{A.5}$$

# B Conventional EHD contact models

Conventional EHD contact models are used in different parts of this work to compute a reference solution. In the following, these models are summarized.

The conventional EHD line contact model is summarized in Section B.1. It is used for the shock loading example problem to study time integration procedures, see Section 4.1.

The conventional EHD axisymmetric point contact model is summarized in Section B.2. It is used for the problem of the lubricated impact to allow comparison with the extended EHD contact model, see Section 4.3.

## B.1 EHD line contact model

The EHD contact model commonly consists of the Reynolds Equation to model fluid flow, the elastic half-space approach to model elastic deformation and the load balance balancing the contact force with the external load. The example problem in Section 4.1 uses this modeling approach. Here, it is referred to as the conventional EHD contact model. Its derivation can be found for example in [66, 128].

The Reynolds Equation considering entraining motion and squeeze motion has the following form

$$\frac{\partial}{\partial x}\left(\frac{\rho h^3}{12\eta}\frac{\partial p}{\partial x}\right) - u_{\mathrm{m}}\frac{\partial \rho h}{\partial x} - \frac{\partial \rho h}{\partial t} - \gamma_{\mathrm{pen}}\min(p, 0) = 0, \tag{B.1}$$

with the pressure $p$, the density $\rho$, the viscosity $\eta$, the gap height $h$, the mean surface velocity $u_{\mathrm{m}}$, the spatial coordinate $x$, and time $t$. The penalty factor $\gamma_{\mathrm{pen}}$ is some arbitrarily large number. Thereby, the last term "penalizes" a negative pressure and thus enforces the Swift-Stieber conditions. The first term of Reynolds Equation is referred to as the Poiseuille term, the second term is referred to as Couette term and the time-dependent term is referred to as the squeeze term.

For the line contact problem classically vanishing pressure boundary conditions are applied on both sides of the domain, i.e.,

$$\begin{aligned} p(x_{\mathrm{L}}, t) &= 0, \ \forall t \geq t_0, \text{ and} \\ p(x_{\mathrm{R}}, t) &= 0, \ \forall t \geq t_0. \end{aligned} \tag{B.2}$$

Additionally, an initial pressure distribution is prescribed as $p(x, 0) = p_0(x)$.

The deformed gap height is given by

$$h(t,\ x) = \frac{x^2}{2\,R_{\text{red}}} + \delta(t,\ x) + h_{00}(t), \tag{B.3}$$

with the undeformed gap height, which is expressed as $x^2/(2\,R_{\text{red}})$, the gap displacement $\delta(t,\ x)$ and the rigid body displacement $h_{00}(t)$. Different from the extended model in Chapter 3, in the conventional method the deformation is approximated using the elastic half-space approach [45],

$$\delta(t,\ x) = -\frac{2}{\pi E'} \int_{\Omega} p(t,\ x') \ln\left((x - x')^2\right) \mathrm{d}x' + C(t). \tag{B.4}$$

Since the rigid body motion is not taken into account in this model, the integration constant $C(t)$ is generally not explicitly determined. Hence, the final unknown, a combination of the rigid body displacement $h_{00}$ and the integration constant $C(t)$, can be computed via the load balance

$$w(t) = \int_{\Omega} p(t,\ x')\,\mathrm{d}x. \tag{B.5}$$

As in Subsection 2.1.3 the problem is transformed into the dimensionless form. For the entraining motion contact, however, the Hertzian pressure $p_H$ and the Hertzian half-width $b$ are computed based on the reference force $w_{\text{ref}}$ as

$$p_H = \frac{2\,w_{\text{ref}}}{\pi\,b} \text{ and } b = \sqrt{\frac{8\,w_{\text{ref}} R_{\text{red}}}{E'\pi}}. \tag{B.6}$$

The reference force is used to describe the time-dependent force as

$$w(t) = w_{\text{ref}}\bar{w}(t). \tag{B.7}$$

With the transformation relations (2.19), equations (B.1) through (B.5) are transformed into dimensionless form

$$\frac{\partial}{\partial X}\left(\frac{\bar{\rho} H^3}{\bar{\eta}\bar{\lambda}} \frac{\partial P}{\partial X}\right) - \frac{\partial \bar{\rho} H}{\partial X} - \frac{\partial \bar{\rho} H}{\partial T} - \bar{\gamma}_{\text{pen}} \min(P, 0) = 0, \tag{B.8}$$

$$\Delta + \frac{1}{2\pi} \int_{\Omega} P(X',\ T) \ln\left((X - X')^2\right) \mathrm{d}X' = 0, \text{ and} \tag{B.9}$$

$$\int_{\Omega} P(X,\ T)\,\mathrm{d}X - \frac{\pi}{2}\bar{w}(T) = 0, \tag{B.10}$$

with $\lambda = \left(12u_{\text{ref}}\eta_0 R_{\text{red}}^2\right) / \left(b^3 p_H\right)$ and

$$H(T, X) = \frac{X^2}{2} + \overline{\delta}(T, X) + H_0(T). \tag{B.11}$$

The Reynolds Equation is discretized in space using second order finite difference scheme according to [128]. The temporal discretization is carried out using Alexander's second-order method, see Section 2.4. After discretization, the resulting system of equations is solved using the Newton-Raphson method, see Section 3.6.

## B.2 Conventional structural model

In Section 4.3 the lubricated impact problem is investigated. In the scope of that investigation comparisons of the extended EHD contact model with the conventional EHD contact model are carried out. The conventional EHD contact model for the lubricated impact problem uses the same fluid model and same nondimensionalization as given in Section 2.1. Hence, the main difference between the extended EHD contact model and the conventional EHD contact model is the structural model, which is used to compute the gap height $h$. The structural model used in this context is referred to as the conventional structural model throughout this work.

In the conventional structural model, the deformation is approximated using the half-space approach and the dynamic behavior is approximated as the motion of a mass point. In the following, the structural model for the axisymmetric point contact is given. To this end, polar coordinates are used. For the sake of consistency, however, the radial coordinate is called $x$. Analogous to above, the deformed gap height is given as

$$h(t, x) = \frac{x^2}{2\, R_{\text{red}}} + \delta(t, x) + h_{00}(t), \tag{B.12}$$

with the undeformed gap height $x^2/(2\, R_{\text{red}})$, the gap displacements $\delta$ and the rigid body displacement $h_{00}$. The rigid body displacement $h_{00}$ can be computed from the equation of motion of the sphere

$$\frac{\partial^2 h_{00}}{\partial t^2} = \frac{2\pi}{m_0} \int_0^\infty p(t, x)\, x\, \mathrm{d}x - g, \tag{B.13}$$

with the mass of the sphere $m_0$ and the acceleration acting on the sphere $g$.

The gap displacements are computed in dimensionless form. After spatial discretization, the dimensionless gap displacements are listed in the vector $\bar{\boldsymbol{\delta}} \in \mathbb{R}^{n_F}$, where $n_F$ is the number of fluid grid points. In the conventional structural model the gap displacements are computed using the compliance matrix $\mathbf{C}_{\mathrm{HST}} \in \mathbb{R}^{n_F \times n_F}$. The coefficients of the compliance matrix are given in [138]. However, if a finite element model is given, a constraint mode matrix can be used to construct a compliance matrix, see Subsection 2.3.4. With the precomputed compliance matrix the dimensionless gap displacements are computed as

$$\bar{\boldsymbol{\delta}} = \mathbf{C}_{\mathrm{HST}} \boldsymbol{P}, \tag{B.14}$$

where the vector $\boldsymbol{P} \in \mathbb{R}^{n_F}$ contains the dimensionless pressures.

# C Time integration of boundary conditions and parameters

Typically, as the temporal behavior of the boundary conditions or parameters is known, they are not considered in the time integration procedure. Hence, they are not considered in the step size control approach. However, as discussed in Subsection 4.1.3, considering the temporal evolution of boundary conditions and parameter functions during step size selection, can be advantageous.

The following describes a procedure for calculating the local integration error of a predefined function. This procedure can be applied to functions of parameters and boundary conditions.[1] In the following, the procedure is applied to the load function of the example problem in Section 4.1.

The load function $\bar{w}(T)$ is given via Eq. (4.1). To compute the local integration error of the load function, the following steps are taken:

1. Differentiate the analytical function numerically using second-order complex step differentiation

$$\frac{\partial \bar{w}(T)}{\partial T} \approx \frac{\operatorname{Im}\left(\bar{w}\left(T + \mathrm{i}h\right)\right)}{h}, \tag{C.1}$$

   which evaluates the function at the imaginary argument $T + \mathrm{i}h$, with the step size $h$.[2] Different from finite difference based methods, complex step differentiation allows for very small step sizes and is limited rather by errors due to underflow rather than due to loss of significance, see [96]. Here, the machine epsilon for double precision is chosen as the step size, i.e., $h = \epsilon = 2.2 \cdot 10^{-16}$.

[1] The proper time integration of the boundary conditions is crucial for singly diagonal implicit Runge-Kutta methods. Time-dependent boundary conditions should be integrated at the stage values. Prescribing the analytical value of the boundary conditions at the stage values might cause order reductions, cf. [105, 112].

[2] The numerically cheapest and most accurate way to compute the derivative would be to calculate an analytical derivative. However, this approach aims to be flexible to work with different functions. Therefore, the derivative is approximated numerically. Complex step differentiation offers a numerically cheap and step size insensitive way to obtain a high accuracy approximation of the derivative. The interested reader may refer to [91, 96, 118] for more information. The error component introduced by the numerical differentiation, is considered negligible for this application.

2. Integrate the numerical derivative given by Eq. (C.1) numerically, using the same time integration scheme as is used for the problem. The integration yields an approximated value of the load function $\bar{w}_{j+1}$ at time $T_{j+1}$.

3. The local integration error for the load function can then be approximated as

$$\bar{w}^{\text{err}}\left(T_{j+1}\right) = \frac{\left\|\bar{w}\left(T_{j+1}\right) - \bar{w}_{j+1}\right\|}{\left\|\bar{w}\left(T_{j+1}\right)\right\|}, \tag{C.2}$$

   where $\bar{w}\left(T_{j+1}\right)$ is the analytical value of the load function at time $T_{j+1}$.

The local integration error of the load function is computed in each time step. If the approximated local integration error $w^{\text{err}}\left(T_{j+1}\right)$ is higher than a prescribed threshold $\epsilon_w$, the time step $\Delta T_j$ is decreased until $w^{\text{err}}\left(T_{j+1}\right)$ falls below the threshold $\epsilon_w$.

# References


[1] R. Alexander. Diagonally Implicit Runge–Kutta Methods for Stiff O.D.E.'s. *SIAM Journal on Numerical Analysis*, 14(6):1006–1021, 1977.

[2] D. Amsallem, M. J. Zahr, and C. Farhat. Nonlinear model order reduction based on local reduced-order bases. *International Journal for Numerical Methods in Engineering*, 92(10):891–916, Dec. 2012.

[3] A. C. Antoulas and D. C. Sorensen. Approximation of large-scale dynamical systems: An overview. *International Journal of Applied Mathematics and Computer Science*, 11(5):1093–1122, 2001.

[4] M. C. C. Bampton and R. R. Craig, Jr. Coupling of substructures for dynamic analyses. *AIAA Journal*, 6(7):1313–1319, July 1968.

[5] D. Bartel. *Simulation von Tribosystemen: Grundlagen und Anwendungen*. Vieweg + Teubner, Wiesbaden, 1st edition, 2010.

[6] C. Barus. Isothermals, isopiestics and isometrics relative to viscosity. *American Journal of Science*, s3-45(266):87–96, Feb. 1893.

[7] K. J. Bathe. *Finite-Elemente-Methoden*. Springer, Berlin, 2nd edition, 2002.

[8] K. J. Bathe and E. L. Wilson. Stability and accuracy analysis of direct integration methods. *Earthquake Engineering & Structural Dynamics*, 1(3):283–291, Jan. 1972.

[9] Belyi, M, Belsky, V, Bajer, A, Kim, M, and Ianculescu, C. Advanced linear dynamics and substructuring capabilities in Abaqus with applications in noise and vibration analysis. In *Proceedings of ISMA*, pages 3427–3438, 2012.

[10] F. P. Bowden and D. Tabor. *The friction and lubrication of solids*. Oxford University Press, Oxford, 1950.

[11] A. Brandt and A. Lubrecht. Multilevel matrix multiplication and fast solution of integral equations. *Journal of Computational Physics*, 90(2):348–370, Oct. 1990.

[12] M. J. Braun and W. M. Hannon. Cavitation formation and modelling for fluid film bearings: a review. *Proceedings of the Institution of Mechanical Engineers, Part J: Journal of Engineering Tribology*, 224(9):839–863, Jan. 2010.

[13] K. Carlberg. Efficient non-linear model reduction via a least-squares Petrov–Galerkin projection and compressive tensor approximations. *International Journal for Numerical Methods in Engineering*, 86(2):155–181, 2011.

[14] K. Carlberg. *Model Reduction of Nonlinear Mechanical Systems via Optimal Projection and Tensor Approximation*. Ph.D. thesis, Stanford University, U.S.A., 2011.

[15] A. Chatterjee. An introduction to the proper orthogonal decomposition. *Current science*, 78(7):808–817, Apr. 2000.

[16] S. Chaturantabut. *Nonlinear model reduction via discrete empirical interpolation*. Ph.D. thesis, Rice University, U.S.A., 2011.

[17] S. Chaturantabut and D. Sorensen. Discrete Empirical Interpolation for nonlinear model reduction. In *Proceedings of the 48h IEEE Conference on Decision and Control (CDC) held jointly with 2009 28th Chinese Control Conference*, pages 4316–4321, 2009.

[18] S. Chaturantabut and D. C. Sorensen. Nonlinear Model Reduction via Discrete Empirical Interpolation. *SIAM Journal on Scientific Computing*, 32(5):2737–2764, Jan. 2010.

[19] S.-H. Chen and H. H. Pan. Guyan reduction. *Communications in applied numerical methods*, 4(4):549–556, 1988.

[20] R. J. Chittenden, D. Dowson, J. F. Dunn, and C. M. Taylor. A Theoretical Analysis of the Isothermal Elastohydrodynamic Lubrication of Concentrated Contacts. I. Direction of Lubricant Entrainment Coincident with the Major Axis of the Hertzian Contact Ellipse. *Proceedings of the Royal Society A: Mathematical, Physical and Engineering Sciences*, 397(1813):245–269, Feb. 1985.

[21] H. Christensen. The oil film in a closing gap. *Proceedings of the Royal Society of London A: Mathematical, Physical and Engineering Sciences*, 266(1326):312–328, 1962.

[22] H. Christensen. Elastohydrodynamic theory of spherical bodies in normal approach. *Journal of Lubrication Technology*, 92(1):145–153, 1970.

[23] J. Chung and G. M. Hulbert. A time integration algorithm for structural dynamics with improved numerical dissipation: the generalized-$\alpha$ method. *Journal of applied mechanics*, 60(2):371–375, 1993.

[24] H. D. Conway and H. C. Lee. Impact of a Lubricated Surface by a Sphere. *Journal of Lubrication Technology*, 97(4):613–615, 1975.

[25] R. Craig and C.-J. Chang. Free-interface methods of substructure coupling for dynamic analysis. *AIAA Journal*, 14(11):1633–1635, Nov. 1976.

[26] R. Craig, Jr. Coupling of substructures for dynamic analyses: An overview. In *Proceedings of the 41st AIAA/ASME/ASCE/AHS/ASC Structures, Structural Dynamics and Materials Conference and Exhibit*, AIAA Paper 2000-1573, pages 1–12, Apr. 2000.

[27] R. R. Craig Jr. *Structural dynamics: an introduction to computer methods.* Wiley, New York, 1981.

[28] R. R. Craig Jr. A review of time-domain and frequency-domain component mode synthesis method. In *Proceedings of the Joint ASCE/ASME Mechanics Conference*, pages 1–30. ASME, 1985.

[29] A. Czekanski, N. El-Abbasi, and S. A. Meguid. Optimal time integration parameters for elastodynamic contact problems. *Communications in Numerical Methods in Engineering*, 17(6):379–384, June 2001.

[30] Dassault Systèmes. *ABAQUS 6.14: ABAQUS/Standard Analysis User's Manual.* 2014.

[31] Dassault Systèmes. *ABAQUS 6.14: Theory Manual.* 2014.

[32] S. Diebels, P. Ellsiepen, and W. Ehlers. Error-controlled Runge-Kutta time integration of a viscoplastic hybrid two-phase model. *Technische Mechanik*, 19:19–27, 1999.

[33] D. Dowson and P. Ehret. Past, present and future studies in elastohydrodynamics. *Proceedings of the Institution of Mechanical Engineers, Part J: Journal of Engineering Tribology*, 213(5):317–333, May 1999.

[34] D. Dowson and G. R. Higginson. A Numerical Solution to the Elasto-Hydrodynamic Problem. *Journal of Mechanical Engineering Science*, 1(1):6–15, June 1959.

[35] D. Dowson and G. R. Higginson. *Elasto-hydrodynamic lubrication: the fundamentals of roller and gear lubrication.* Pergamon Press, Oxford, 1st edition, 1966.

[36] D. Dowson and G. R. Higginson. Reflections on Early Studies of Elasto-Hydrodynamic Lubrication. In R. W. Snidle and H. P. Evans, editors, *IUTAM Symposium on Elastohydrodynamics and Micro-elastohydrodynamics*, volume 134, pages 3–21. Springer Netherlands, 2006.

[37] D. Dowson and D. A. Jones. Lubricant entrapment between approaching elastic solids. *Nature*, 214:947–948, 1967.

[38] D. Dowson and D. Wang. An analysis of the normal bouncing of a solid elastic ball on an oily plate. *Wear*, 179(1-2):29–37, 1994.

[39] P. Ellsiepen. *Zeit- und ortsadaptive Verfahren angewandt auf Mehrphasenprobleme poröser Medien.* Ph.D. thesis, University of Stuttgart, Germany, July 1999.

[40] P. Ellsiepen and S. Hartmann. Remarks on the interpretation of current non-linear finite element analyses as differential–algebraic equations. *International Journal for Numerical Methods in Engineering*, 51(6):679–707, June 2001.

[41] A. Ertel. Hydrodynamic lubrication based on new principles. *Akad. Nauk. SSSR, Prikladnaya Matematika i Mekhanika*, 3(2):41–52, 1939.

[42] H. P. Evans and T. G. Hughes. Evaluation of deflection in semi-infinite bodies by a differential method. *Proceedings of the Institution of Mechanical Engineers, Part C: Journal of Mechanical Engineering Science*, 214(4):563–584, Apr. 2000.

[43] H. P. Evans and R. W. Snidle. The Elastohydrodynamic Lubrication of Point Contacts at Heavy Loads. *Proceedings of the Royal Society A: Mathematical, Physical and Engineering Sciences*, 382(1782):183–199, July 1982.

[44] D. J. Ewins. *Modal testing: theory, practice, and application*. Research Studies Press Ltd., 2nd edition, 2000.

[45] A. Flamant. Sur la répartition des pressions dans un solide rectangulaire chargé transversalement. *CR Acad. Sci. Paris*, 114:1465–1468, 1892.

[46] W. Goldsmith. *Impact: the theory and physical behaviour of colliding solids*. Dover Publications, Mineola, N.Y., 2001.

[47] C. E. Goodyer. *Adaptive numerical methods for elastohydrodynamic lubrication*. Ph.D. thesis, The University of Leeds, England, 2001.

[48] C. E. Goodyer and M. Berzins. Adaptive Timestepping for Elastohydrodynamic Lubrication Solvers. *SIAM Journal on Scientific Computing*, 28(2):626–650, Jan. 2006.

[49] G. L. Goudreau and R. L. Taylor. Evaluation of numerical integration methods in elastodynamics. *Computer Methods in Applied Mechanics and Engineering*, 2(1):69–97, Feb. 1973.

[50] D. Gross, W. Hauger, J. Schröder, and W. A. Wall. *Technische Mechanik 2*. Springer, Berlin, Heidelberg, 2009.

[51] A. Grubin. Fundamentals of the Hydrodynamic Theory of Lubrication of Heavily Loaded Cylindrical Surfaces. *Central Scientific Institute for Technology and Mechanical Engineering*, 30:115–166, 1949.

[52] F. Guo, M. Kaneta, J. Wang, H. Nishikawa, and P. Yang. Occurrence of a Noncentral Dimple in Squeezing EHL Contacts. *Journal of Tribology*, 128(3):632–640, 2006.

[53] K. Gustafsson. Control-theoretic Techniques for Stepsize Selection in Implicit Runge-Kutta Methods. *ACM Trans. Math. Softw.*, 20(4):496–517, 1994.

[54] R. J. Guyan. Reduction of stiffness and mass matrices. *AIAA Journal*, 3(2):380–380, Feb. 1965.

[55] W. Habchi. *A full-system finite element approach to elastohydrodynamic lubrication problems: application to ultra-low-viscosity fluids*. Ph.D. thesis, University of Lyon, France, 2008.

[56] W. Habchi. Reduced order finite element model for elastohydrodynamic lubrication: Circular contacts. *Tribology International*, 71:98–108, 2014.

[57] W. Habchi and J. Issa. A Reduced Full-System Finite Element Approach to the Solution of EHL Problems: Line Contacts. In *Proceedings of the STLE/ASME 2010 International Joint Tribology Conference*, pages 95–97, Jan. 2010.

[58] W. Habchi and J. Issa. Fast and reduced full-system finite element solution of elastohydrodynamic lubrication problems: Line contacts. *Advances in Engineering Software*, 56:51–62, Feb. 2013.

[59] W. Habchi and J. S. Issa. An Exact and General Model Order Reduction Technique for the Finite Element Solution of Elastohydrodynamic Lubrication Problems. *Journal of Tribology*, 139(5):051501–051501–19, May 2017.

[60] W. Habchi, D. Eyheramendy, P. Vergne, and G. Morales-Espejel. A Full-System Approach of the Elastohydrodynamic Line/Point Contact Problem. *Journal of Tribology*, 130(2): 021501–021501–10, 2008.

[61] W. Habchi, D. Eyheramendy, P. Vergne, and G. Morales-Espejel. Stabilized fully-coupled finite elements for elastohydrodynamic lubrication problems. *Advances in Engineering Software*, 46(1):4–18, Apr. 2012.

[62] E. Hairer and G. Wanner. *Solving Ordinary Differential Equations II*, volume 14 of *Springer Series in Computational Mathematics*. Springer, Berlin, Heidelberg, 1996.

[63] E. Hairer, S. P. Norsett, and G. Wanner. *Solving Ordinary Differential Equations I*, volume 8 of *Springer Series in Computational Mathematics*. Springer, Berlin, Heidelberg, 1993.

[64] B. J. Hamrock and D. Dowson. Isothermal Elastohydrodynamic Lubrication of Point Contacts: Part 1—Theoretical Formulation. *Journal of Lubrication Technology*, 98(2): 223–228, 1976.

[65] B. J. Hamrock and B. O. Jacobson. Elastohydrodynamic Lubrication of Line Contacts. *A S L E Transactions*, 27(4):275–287, Jan. 1984.

[66] B. J. Hamrock, S. R. Schmid, and B. O. Jacobson. *Fundamentals of Fluid Film Lubrication*. CRC Press, Baton Rouge, Mar. 2004.

[67] S. Hartmann. *Finite-Elemente-Berechnung inelastischer Kontinua: Interpretation als Algebro-Differentialgleichungssysteme*. Habilitationsschrift, University of Kassel, Germany, 2003.

[68] P. Haupt. *Continuum Mechanics and Theory of Materials*. Springer, Berlin, 2002.

[69] K. Herrebrugh. Elastohydrodynamic squeeze films between two cylinders in normal approach. *Journal of Lubrication Technology*, 92(2):292–301, 1970.

[70] H. Hertz. Über die Berührung fester elastischer Körper. *Journal für die reine und angewandte Mathematik*, 1882(92):156–171, 1881.

[71] H. M. Hilber, T. J. R. Hughes, and R. L. Taylor. Improved numerical dissipation for time integration algorithms in structural dynamics. *Earthquake Engineering & Structural Dynamics*, 5(3):283–292, July 1977.

[72] M. J. A. Holmes, H. P. Evans, T. G. Hughes, and R. W. Snidle. Transient elastohydrodynamic point contact analysis using a new coupled differential deflection method Part 1: Theory and validation. *Proceedings of the Institution of Mechanical Engineers, Part J: Journal of Engineering Tribology*, 217(4):289–304, Apr. 2003.

[73] M. J. A. Holmes, H. P. Evans, T. G. Hughes, and R. W. Snidle. Transient elastohydrodynamic point contact analysis using a new coupled differential deflection method Part 2: Results. *Proceedings of the Institution of Mechanical Engineers, Part J: Journal of Engineering Tribology*, 217(4):305–322, Apr. 2003.

[74] L. G. Houpert and B. J. Hamrock. Fast approach for calculating film thicknesses and pressures in elastohydrodynamically lubricated contacts at high loads. *Journal of Tribology*, 108(3):411–419, 1986.

[75] T. G. Hughes, C. D. Elcoate, and H. P. Evans. Coupled solution of the elastohydrodynamic line contact problem using a differential deflection method. *Proceedings of the Institution of Mechanical Engineers, Part C: Journal of Mechanical Engineering Science*, 214(4): 585–598, Apr. 2000.

[76] T. J. R. Hughes and T. Hughes. *The Finite Element Method: Linear Static and Dynamic Finite Element Analysis*. Dover Publications, Mineola, N.Y., 2000.

[77] W. C. Hurty. Dynamic analysis of structural systems using component modes. *AIAA Journal*, 3(4):678–685, Apr. 1965.

[78] B. Jakobsson and L. Floberg. The finite journal bearing, considering vaporization. Chalmers Tekniska Högskolas handlingar. Gumperts Förlag, Göteborg, 1957.

[79] K. L. Johnson. *Contact mechanics*. Cambridge University Press, Cambridge, 1985.

[80] M. Kaneta, S. Ozaki, H. Nishikawa, and F. Guo. Effects of impact loads on point contact elastohydrodynamic lubrication films. *Proceedings of the Institution of Mechanical Engineers, Part J: Journal of Engineering Tribology*, 221(3):271–278, Jan. 2007.

[81] M. Kaneta, F. Guo, and J. Wang. Impact Micro-Elastohydrodynamics in Point Contacts. *Journal of Tribology*, 133(3):031503–031503–9, July 2011.

[82] M. Kaneta, H. Nishikawa, M. Mizui, and F. Guo. Impact elastohydrodynamics in point contacts. *Proceedings of the Institution of Mechanical Engineers, Part J: Journal of Engineering Tribology*, 225(1):1–12, 2011.

[83] M. Kim, V. Belsky, and M. Belyi. Substructure generation using automated multilevel substructuring, May 2013. U.S. Classification 703/1; International Classification G06F17/50; Cooperative Classification G06F17/5018, G06F17/5095.

[84] J. Koreck and O. von Estorff. Reduced order structural models for the calculation of wet contact forces due to impacts in hydraulic valves. *Meccanica*, 50(5):1387–1401, May 2015.

[85] R. Larsson and E. Höglund. Numerical Simulation of a Ball Impacting and Rebounding a Lubricated Surface. *Journal of Tribology*, 117(1):94–102, Jan. 1995.

[86] K. M. Lee and H. S. Cheng. The Pressure and Deformation Profiles Between Two Normally Approaching Lubricated Cylinders. *Journal of Lubrication Technology*, 95(3): 308–317, July 1973.

[87] A. A. Lubrecht. *The numerical solution of the elastohydrodynamically lubricated line and point contact problem, using multigrid techniques*. Ph.D. thesis, University of Twente, Netherlands, 1987.

[88] A. A. Lubrecht, W. E. Ten Napel, and R. Bosma. Multigrid, an alternative method for calculating film thickness and pressure profiles in elastohydrodynamically lubricated line contacts. *Journal of tribology*, 108(4):551–556, 1986.

[89] A. A. Lubrecht, W. E. Ten Napel, and R. Bosma. Multigrid, an alternative method of solution for two-dimensional elastohydrodynamically lubricated point contact calculations. *Journal of tribology*, 109(3):437–443, 1987.

[90] P. M. Lugt and G. E. Morales-Espejel. A Review of Elasto-Hydrodynamic Lubrication Theory. *Tribology Transactions*, 54(3):470–496, 2011.

[91] J. N. Lyness and C. B. Moler. Numerical Differentiation of Analytic Functions. *SIAM Journal on Numerical Analysis*, 4(2):202–210, June 1967.

[92] R. H. MacNeal. A hybrid method of component mode synthesis. *Computers & Structures*, 1(4):581–601, 1971.

[93] D. Maier. *On the Use of Model Order Reduction Techniques for the Elastohydrodynamic Contact Problem*. Ph.D. thesis, Karlsruhe Institute of Technology, Germany, 2015.

[94] D. Maier, C. Hager, H. Hetzler, N. Fillot, P. Vergne, D. Dureisseix, and W. Seemann. Fast Solution of Transient Elastohydrodynamic Line Contact Problems Using the Trajectory Piecewise Linear Approach. *Journal of Tribology*, 138(1):011502–011502–9, Aug. 2015.

[95] D. Maier, C. Hager, H. Hetzler, N. Fillot, P. Vergne, D. Dureisseix, and W. Seemann. A nonlinear model order reduction approach to the elastohydrodynamic problem. *Tribology International*, 82, Part B:484–492, 2015.

[96] J. R. R. A. Martins, P. Sturdza, and J. J. Alonso. The complex-step derivative approximation. *ACM Transactions on Mathematical Software*, 29(3):245–262, Sept. 2003.

[97] D. P. Mok. *Partitionierte Lösungsansätze in der Strukturdynamik und der Fluid-Struktur-Interaktion*. Ph.D. thesis, Universität Stuttgart, Germany, 2001.

[98] H. Nishikawa, H. Miyazaki, M. Kaneta, and F. Guo. Effects of two-stage impact load on point contact elastohydrodynamic lubrication films. *Proceedings of the Institution of Mechanical Engineers, Part J: Journal of Engineering Tribology*, 222(7):807–814, Nov. 2008.

[99] C. C. Ober and J. N. Shadid. Studies on the accuracy of time-integration methods for the radiation–diffusion equations. *Journal of Computational Physics*, 195(2):743–772, Apr. 2004.

[100] K. P. Oh. The Numerical Solution of Dynamically Loaded Elastohydrodynamic Contact as a Nonlinear Complementarity Problem. *Journal of Tribology*, 106(1):88–95, Jan. 1984.

[101] H. Okamura. A contribution to the numerical analysis of isothermal elastohydrodynamic lubrication. In *Proceedings of the Ninth Leeds-Lyon Symposium on Tribology*, pages 313–320, 1982.

[102] K.-O. Olsson. Cavitation in dynamically loaded bearings. Chalmers tekniska Högskolas Handlingar. Scandinavian University Books, Göteborg, 1965.

[103] B. Owren and H. H. Simonsen. Alternative integration methods for problems in structural dynamics. *Computer Methods in Applied Mechanics and Engineering*, 122(1-2):1–10, Apr. 1995.

[104] A. I. Petrusevich. Fundamental conclusions from the contact-hydrodynamic Theory of Lubrication. *Izv. Akad. Nauk SSR. Otd. Tekh. Nauk*, 2:209–233, 1951.

[105] A. Prothero and A. Robinson. On the stability and accuracy of one-step methods for solving stiff systems of ordinary differential equations. *Mathematics of Computation*, 28(125):145–162, 1974.

[106] Z.-Q. Qu. *Model Order Reduction Techniques*. Springer, London, 2004.

[107] A. P. Ranger, C. M. M. Ettles, and A. Cameron. The Solution of the Point Contact Elasto-Hydrodynamic Problem. *Proceedings of the Royal Society A: Mathematical, Physical and Engineering Sciences*, 346(1645):227–244, Oct. 1975.

[108] M. Rewienski and J. White. A trajectory piecewise-linear approach to model order reduction and fast simulation of nonlinear circuits and micromachined devices. *IEEE Transactions on Computer-Aided Design of Integrated Circuits and Systems*, 22(2): 155–170, Feb. 2003.

[109] M. J. Rewienski. *A trajectory piecewise-linear approach to model order reduction of nonlinear dynamical systems*. Ph.D. thesis, Massachusetts Institute of Technology, U.S.A., 2003.

[110] O. Reynolds. On the Theory of Lubrication and Its Application to Mr. Beauchamp Tower's Experiments, Including an Experimental Determination of the Viscosity of Olive Oil. *Proceedings of the Royal Society of London*, 40(242-245):191–203, Jan. 1886.

[111] C. J. A. Roelands. *Correlational aspects of the viscosity-temperature-pressure relationship of lubricating oils*. Ph.D. thesis, Delft University of Technology, Netherlands, 1966.

[112] S. Rothe, J. H. Schmidt, and S. Hartmann. Analytical and numerical treatment of electro-thermo-mechanical coupling. *Archive of Applied Mechanics*, 85(9-10):1245–1264, Sept. 2015.

[113] S. Rubin. Improved Component-Mode Representation for Structural Dynamic Analysis. *AIAA Journal*, 13(8):995–1006, Aug. 1975.

[114] M. M. A. Safa and R. Gohar. Pressure distribution under a ball impacting a thin lubricant layer. *Journal of tribology*, 108(3):372–376, 1986.

[115] M. Sakamoto, H. Nishikawa, and M. Kaneta. Behaviour of point contact EHL films under pulsating loads. *Tribology Series*, 43:391–399, 2003.

[116] G. Söderlind. Time-step selection algorithms: Adaptivity, control, and signal processing. *Applied Numerical Mathematics*, 56(3-4):488–502, Mar. 2006.

[117] H. A. Spikes. Sixty years of EHL. *Lubrication Science*, 18(4):265–291, 2006.

[118] W. Squire and G. Trapp. Using Complex Variables to Estimate Derivatives of Real Functions. *SIAM Review*, 40(1):110–112, Jan. 1998.

[119] R. R. Stephenson and J. F. Osterle. A Direct Solution of the Elasto-Hydrodynamic Lubrication Problem. *A S L E Transactions*, 5(2):365–374, Jan. 1962.

[120] W. Stieber. *Das Schwimmlager: Hydrodynamische Theorie des Gleitlagers*. VDI Verlag, Berlin, 1933.

[121] W. J. Stronge. *Impact Mechanics*. Cambridge University Press, Cambridge, Mar. 2004.

[122] H. W. Swift. The stability of lubricating films in journal bearings. *Minutes of the Proceedings of the Institution of Civil Engineers*, 233(1932):267–288, Jan. 1932.

[123] A. Z. Szeri. *Fluid film lubrication*. Cambridge University Press, Cambridge, 2010.

[124] C. Venner and A. Lubrecht. Amplitude Reduction of Non-Isotropic Harmonic Patterns in Circular EHL Contacts, under Pure Rolling. In *Lubrication at the Frontier, Proceedings of the 25th Leeds-Lyon Symposium on Tribology*, volume 36 of *Tribology Series*, pages 151–162. Elsevier, 1999.

[125] C. Venner, J. Wang, and A. Lubrecht. Central film thickness in EHL Point Contacts under Pure Impact Revisited. *Tribology International*, 100:1–6, 2016.

[126] C. H. Venner. *Multilevel solution of the EHL line and point contact problems*. Ph.D. thesis, University of Twente, Netherlands, 1991.

[127] C. H. Venner and A. A. Lubrecht. Transient Analysis of Surface Features in an EHL Line Contact in the Case of Sliding. *Journal of Tribology*, 116(2):186–193, 1994.

[128] C. H. Venner and A. A. Lubrecht. *Multi-level Methods in Lubrication*. Elsevier Science, Amsterdam, July 2000.

[129] J. Wang, C. H. Venner, and A. A. Lubrecht. Central film thickness prediction for line contacts under pure impact. *Tribology International*, 66:203–207, 2013.

[130] J. Wang, N. Wang, P. Yang, M. Kaneta, and A. A. Lubrecht. A theoretical simulation of thermal elastohydrodynamic lubrication for a Newtonian fluid in impact motion. *Tribology International*, 67:116–123, Nov. 2013.

[131] J. Wang, P. Yang, A. Lubrecht, and M. Kaneta. Numerical investigation of thermal EHL in elliptical contact under impact motion. *Proceedings of the Institution of Mechanical Engineers, Part J: Journal of Engineering Tribology*, 229(9):1125–1131, Sept. 2015.

[132] B. Watremetz, F. Colin, C. Venner, and A. Lubrecht. Optimum time step in a transient EHL contact. In *Transient Processes in Tribology*, volume 43 of *Tribology Series*, pages 591–597. Elsevier, 2003.

[133] B. Wiegert. *Nichtlineare Schwingungen von Systemen mit elastohydrodynamischen Linienkontakten*. Ph.D. thesis, Karlsruhe Institute of Technology, Germany, 2015.

[134] Y. H. Wijnant. *Contact dynamics in the field of elastohydrodynamic lubrication*. Ph.D. thesis, University of Twente, Netherlands, 1998.

[135] Y. H. Wijnant and C. H. Venner. Analysis of an EHL circular contact incorporating rolling element vibration. In *Proceedings of the 23rd Leeds-Lyon Symposium on Tribology*, volume 32 of *Tribology Series*, pages 445–456. 1997.

[136] Y. H. Wijnant, J. A. Wensing, and G. C. Nijen. The influence of lubrication on the dynamic behaviour of ball bearings. *Journal of Sound and Vibration*, 222(4):579–596, 1999.

[137] S. Wu. A penalty formulation and numerical approximation of the Reynolds-Hertz problem of elastohydrodynamic lubrication. *International Journal of Engineering Science*, 24(6):1001–1013, Jan. 1986.

[138] P. Yang and S. Wen. Pure squeeze action in an isothermal elastohydrodynamically lubricated spherical conjunction part 1. Theory and dynamic load results. *Wear*, 142(1): 1–16, 1991.

[139] F. Zhang, editor. *The Schur Complement and Its Applications*, volume 4 of *Numerical Methods and Algorithms*. Springer, New York, 2005.

[140] P. Ziegler. Avoiding Unphysical Vibrations Caused by Statically Correct Reduction of Elastic Multibody Systems. Presentation held at the 8th ECCOMAS Thematic Conference on Multibody Dynamic, Prague, Czech Republic, June 2017.

[141] P. Ziegler and P. Holzwarth. Discussion about spurious oscillations [in person], July 2016.

[142] O. Zienkiewicz, R. Taylor, and J. Zhu. *The Finite Element Method: Its Basis and Fundamentals*. Elsevier Butterworth-Heinemann, Oxford, 6th edition, 2005.

# List of student works


[S1] S. Arul. *Impact of a cylinder on a lubricated plate considering elasticity and inertial effects*, Master's Thesis, University of Stuttgart, Germany, 2017

[S2] W. Du. *Impact of a Cylinder on a Lubricated Plate considering Elasticity and Thermal Effects*, Master's Thesis, RWTH Aachen University, Germany, 2016

[S3] M. Meyer. *Weiterentwicklung eines Modells zur Fluid-Struktur-Kopplung in geschmierten Kontakten*, Master's Thesis, Hannover University of Applied Sciences and Arts, Germany, 2015

[S4] A. Patel. *An extension for a fluid-structure-interaction model considering structural dynamics and thermal effects in lubricated contacts*, Master's Thesis, Braunschweig University of Technology, Germany, 2017

[S5] M. Pries. *Untersuchungen zur Gültigkeit vereinfachter Fluidmodelle beim Stoß zweier Festkörper getrennt durch ein Fluid*, Master's Thesis, University of Stuttgart, Germany, 2016

[S6] J. H. Schmidt. *Entwicklung eines Modells zur Fluid-Struktur-Kopplung in einem elastohydrodynamischen Kontakt*, Diploma Thesis, Clausthal University of Technology, Germany, 2014

KARLSRUHER INSTITUT FÜR TECHNOLOGIE (KIT)
SCHRIFTENREIHE DES INSTITUTS FÜR TECHNISCHE MECHANIK (ITM)

**ISSN 1614-3914**

**Band 1** **Marcus Simon**
Zur Stabilität dynamischer Systeme mit stochastischer Anregung. 2004
ISBN 3-937300-13-9

**Band 2** **Clemens Reitze**
Closed Loop, Entwicklungsplattform für mechatronische Fahrdynamikregelsysteme. 2004
ISBN 3-937300-19-8

**Band 3** **Martin Georg Cichon**
Zum Einfluß stochastischer Anregungen auf mechanische Systeme. 2006
ISBN 3-86644-003-0

**Band 4** **Rainer Keppler**
Zur Modellierung und Simulation von Mehrkörpersystemen unter Berücksichtigung von Greifkontakt bei Robotern. 2007
ISBN 978-3-86644-092-0

**Band 5** **Bernd Waltersberger**
Strukturdynamik mit ein- und zweiseitigen Bindungen aufgrund reibungsbehafteter Kontakte. 2007
ISBN 978-3-86644-153-8

**Band 6** **Rüdiger Benz**
Fahrzeugsimulation zur Zuverlässigkeitsabsicherung von karosseriefesten Kfz-Komponenten. 2008
ISBN 978-3-86644-197-2

**Band 7** **Pierre Barthels**
Zur Modellierung, dynamischen Simulation und Schwingungsunterdrückung bei nichtglatten, zeitvarianten Balkensystemen. 2008
ISBN 978-3-86644-217-7

Band 8 **Hartmut Hetzler**
Zur Stabilität von Systemen bewegter Kontinua mit Reibkontakten am Beispiel des Bremsenquietschens. 2008
ISBN 978-3-86644-229-0

Band 9 **Frank Dienerowitz**
Der Helixaktor – Zum Konzept eines vorverwundenen Biegeaktors. 2008
ISBN 978-3-86644-232-0

Band 10 **Christian Rudolf**
Piezoelektrische Self-sensing-Aktoren zur Korrektur statischer Verlagerungen. 2008
ISBN 978-3-86644-267-2

Band 11 **Günther Stelzner**
Zur Modellierung und Simulation biomechanischer Mehrkörpersysteme. 2009
ISBN 978-3-86644-340-2

Band 12 **Christian Wetzel**
Zur probabilistischen Betrachtung von Schienen- und Kraftfahrzeugsystemen unter zufälliger Windanregung. 2010
ISBN 978-3-86644-444-7

Band 13 **Wolfgang Stamm**
Modellierung und Simulation von Mehrkörpersystemen mit flächigen Reibkontakten. 2011
ISBN 978-3-86644-605-2

Band 14 **Felix Fritz**
Modellierung von Wälzlagern als generische Maschinenelemente einer Mehrkörpersimulation. 2011
ISBN 978-3-86644-667-0

Band 15 **Aydin Boyaci**
Zum Stabilitäts- und Bifurkationsverhalten hochtouriger Rotoren in Gleitlagern. 2012
ISBN 978-3-86644-780-6

| | | |
|---|---|---|
| Band 16 | **Rugerri Toni Liong**<br>Application of the cohesive zone model to the analysis of rotors with a transverse crack. 2012<br>ISBN 978-3-86644-791-2 |
| Band 17 | **Ulrich Bittner**<br>Strukturakustische Optimierung von Axialkolbeneinheiten. Modellbildung, Validierung und Topologieoptimierung. 2013<br>ISBN 978-3-86644-938-1 |
| Band 18 | **Alexander Karmazin**<br>Time-efficient Simulation of Surface-excited Guided Lamb Wave Propagation in Composites. 2013<br>ISBN 978-3-86644-935-0 |
| Band 19 | **Heike Vogt**<br>Zum Einfluss von Fahrzeug- und Straßenparametern auf die Ausbildung von Straßenunebenheiten. 2013<br>ISBN 978-3-7315-0023-0 |
| Band 20 | **Laurent Ineichen**<br>Konzeptvergleich zur Bekämpfung der Torsionsschwingungen im Antriebsstrang eines Kraftfahrzeugs. 2013<br>ISBN 978-3-7315-0030-8 |
| Band 21 | **Sietze van Buuren**<br>Modeling and simulation of porous journal bearings in multibody systems. 2013<br>ISBN 978-3-7315-0084-1 |
| Band 22 | **Dominik Kern**<br>Neuartige Drehgelenke für reibungsarme Mechanismen. 2013<br>ISBN 978-3-7315-0103-9 |
| Band 23 | **Nicole Gaus**<br>Zur Ermittlung eines stochastischen Reibwerts und dessen Einfluss auf reibungserregte Schwingungen. 2013<br>ISBN 978-3-7315-0118-3 |
| Band 24 | **Fabian Bauer**<br>Optimierung der Energieeffizienz zweibeiniger Roboter durch elastische Kopplungen. 2014<br>ISBN 978-3-7315-0256-2 |

---

Die Bände sind unter www.ksp.kit.edu als PDF frei verfügbar oder als Druckausgabe zu bestellen.

| | | |
|---|---|
| **Band 25** | **Benedikt Wiegert**<br>Nichtlineare Schwingungen von Systemen<br>mit elastohydrodynamischen Linienkontakten. 2015<br>ISBN 978-3-7315-0350-7 |
| **Band 26** | **Arsenty Tikhomolov**<br>Analytische, numerische und messtechnische<br>Untersuchung der Dynamik von Fahrzeugkupplungen<br>am Beispiel des Trennproblems. 2015<br>ISBN 978-3-7315-0362-0 |
| **Band 27** | **Daniel Maier**<br>On the Use of Model Order Reduction Techniques<br>for the Elastohydrodynamic Contact Problem. 2015<br>ISBN 978-3-7315-0369-9 |
| **Band 28** | **Xiaoyu Zhang**<br>Crosswind stability of vehicles under nonstationary<br>wind excitation. 2015<br>ISBN 978-3-7315-0376-7 |
| **Band 29** | **Jens Deppler**<br>Ein Beitrag zur viskoelastischen Modellierung nichtholonomer<br>Bindungsgleichungen. 2017<br>ISBN 978-3-7315-0548-8 |
| **Band 30** | **Georg Jehle**<br>Zur Modellbildung und Simulation reibungserregter<br>Schwingungen in Pkw-Schaltgetrieben. 2017<br>ISBN 978-3-7315-0668-3 |
| **Band 31** | **Joachim Klima**<br>Lubricant transport towards tribocontact in capillary<br>surface structures. 2018<br>ISBN 978-3-7315-0814-4 |
| **Band 32** | **Gábor Kenderi**<br>Nonparametric identification of nonlinear dynamic systems. 2018<br>ISBN 978-3-7315-0834-2 |

Band 33 **Jan Henrik Schmidt**
An efficient solution procedure for elastohydrodynamic contact problems considering structural dynamics. 2018
ISBN 978-3-7315-0872-4